Pearl

Pearl

Nature's Perfect Gem

Fiona Lindsay Shen

REAKTION BOOKS

To the memory of the indentured and enslaved pearl divers
who for centuries brought us pearls

Published by Reaktion Books Ltd
Unit 32, Waterside
44–48 Wharf Road
London N1 7UX, UK
www.reaktionbooks.co.uk

First published 2022
Copyright © Fiona Lindsay Shen 2022

Printed and bound in India by Replika Press Pvt. Ltd

A catalogue record for this book is available from the British Library

ISBN 978 1 78914 621 9

Contents

Cultured Akoya pearls, Vietnam, 4–4.5 mm.

❧

Sarah Siddons's Necklace

Joshua Reynolds's peerless portrait of the actor Sarah Siddons was an instant sensation at the 1784 Royal Academy exhibition in London. It marked a zenith for both the artist and his sitter. The painting brought viewers to tears, just as Siddons slayed an audience from the stage. The previous year, King George III and Queen Charlotte bawled their way through five of Siddons's performances in January alone. By 1784, 'Siddonimania' gripped the country, igniting a devotion any modern celebrity might envy.[1]

Sarah Siddons as the Tragic Muse enthroned the 28-year-old tragedienne as a cultural icon for the ages. A fashion maven who advised aristocrats on their dress, she wore a stylish but timeless russet robe with billowy white sleeves bracketing a creamy throat. Reynolds, as enamoured as any fan, signed and dated the portrait on her clothing, explaining to his sitter, 'I have resolved to go down to posterity on the hem of your garment.'[2] As it turned out, Siddons was just one of the subjects who secured Reynolds's fame, but what has gone down to posterity are the painted pearls that hang in thick, lustrous hanks, intricately knotted at the dress's neckline. Those pearls reveal the artist's mind at work. They weren't his first choice, but they are the opposite of an afterthought. X-rays show that he first painted Siddons in a pale-coloured crossover bodice with a decorative border.[3] Substituting pearls was a masterstroke that cracked open the painting's meanings and flaunted Reynolds's skill at capturing the soft gleam of the watery gem. They are his radiant encore.

The Dulwich Picture Gallery in London owns another version of the work commissioned in 1789 by the dealer Noël Desenfans. Purportedly, Reynolds agreed to paint a copy in exchange for a Rubens, though his studio assistants executed this later work.[4] The pearls betray the difference. In the original, they twist and glow, each unique gem a reminder of the living animal that made it. In the copy, they hang limp and lifeless, white paint on canvas rather than glistening natural phenomena.

Did such a necklace ever exist? When Reynolds invited Siddons to 'Ascend your undisputed throne, and graciously bestow upon me some good idea of the Tragic Muse', was her neck bowed with the weight of those pearls?[5] Could she have owned such a gravity-defying ornament, one that would surely have required pinning between her breasts? Or, as jewels were borrowed or hired without shame at this time, did she ask a favour of an aristocratic friend?[6] No comparable necklace has ever been unearthed, though it may have been lost in an 1808 fire at Covent Garden Theatre which incinerated Siddons's professional costumes and jewellery, too.[7] Whether Sarah Siddons's necklace ever existed is impossible to know, but its meanings exist, as pearls, though small, are freighted.

Sarah Siddons's beginnings were modest. She was born Sarah Kemble at the Shoulder of Mutton public house in Brecon, Wales. Her parents were undistinguished actors with a travelling theatre company. As a teenager, she served a stint as a maid before marrying the equally undistinguished (though handsome) actor William Siddons. Her hard-scrabble ascent to becoming 'the sublime Siddons', a goddess, divine, 'the stateliest ornament of the public mind', was punctuated by failure and heartache.[8] Financially pressed as the family's main breadwinner, she worked throughout her pregnancies, giving birth to her second child in 1775 halfway through a performance in Gloucester.

And yet, within a few years, she was the confidante of aristocrats including the Duchess of Devonshire and advised socialites on style and fashion. The year before Reynolds's portrait, the royal family

Joshua Reynolds, *Sarah Siddons as the Tragic Muse*, 1783–4, oil on canvas.

retained her as Reader in English for their children. Her social status was unimpeachable. A plain bodice could never do justice to the distance she had travelled from her provincial Welsh birthplace. And so Reynolds changed his mind and gave her pearls – the same gems that graced the throats, wrists and hair of nobility and monarchs. In 1762 the Scottish artist Allan Ramsay had painted a state portrait of Queen Charlotte wearing heavy pearl tassels around her waist and a jaunty knot of pearls on her shoulder, but Siddons's pearls outshine even these.[9]

Other gemstones could have signalled social status, but pearls were a better choice to convey what Siddons meant to the nation. She was, above all else, tragedy personified, revered for her ability to plumb the depths of grief. She elicited not just tears but hysteria and fainting spells. Had she glittered in diamonds or rubies, she would have been far less tragic. This is less to do with diamonds' flashy exuberance than with pearls' long association with sorrow. Myths tell of their creation from the tears of mermaids, angels or goddesses. Even their journey from animal to necklace is melancholy, as the cost of pearls' austere perfection is death. Before cultured pearls' invention in the twentieth century, maybe one mollusc in a thousand contained a perfect white sphere. An extravagant necklace like Siddons wears would have entailed the culling of hundreds of thousands of oysters. Further, pearl fishing often depleted oyster beds, despoiled marine environments and cost the liberty and lives of enslaved divers who were dropped into oceans weighted with stones for a quick descent.

So when Reynolds painted Siddons in pearls, he could count on an audience well-versed in the sadness of pearls. He could also rely on a public equally well-versed in the ongoing tribulations of Siddons's private life. The eighteenth century saw the rise of the celebrity actor whose personal life was avidly consumed.[10] By the time of Reynolds's portrait, Siddons's fourth child had died in infancy, and she had struggled through several illnesses exacerbated by overwork. Exhaustion excluded her from benefit performances in Ireland later that year – for which she was pelted with potatoes in Dublin that July by fans angered by her perceived lack of charity. (She would be publicly accused of stinginess for the rest of her life.) Before the decade was out, she suffered two miscarriages and the death of her six-year-old daughter. The next decade, her teenage daughter died after a traumatic affair with the artist Thomas Lawrence. She lost another daughter five years later, following a disastrous liaison with the same artist. For Siddons, the Tragic Muse was far more than a theatrical role.

Pearls symbolized sadness but so much more. By the Middle Ages in Europe, a firm connection between pearls and chastity made them a perfect symbol for the Virgin Mary. What better gem for such a devoted wife and mother as Sarah Siddons? While refashioning herself as nobility, another hurdle she had to clear was the widespread prejudice against female actors' perceived immorality. Resolutely respectable, avoiding any whiff of scandal, Siddons did much to retrieve the reputation of female actors in the eighteenth century from the realms of disrepute. In fact, her refusal to even display her legs excluded her from comic cross-playing roles, further cementing her career in tragedy.[11]

Significantly, Siddons never appeared in the role of licentious Cleopatra. Twenty-five years before he painted Siddons, Reynolds depicted the famous actor and courtesan Kitty Fisher as Cleopatra decked out in pearls, about to dissolve one the size of her thumb in a goblet of wine. That the same gem could symbolize wasteful luxuriance and chaste discretion is a tribute to its elasticity. This small stone, created by an unprepossessing bivalve, can reflect all our vices and virtues.

What did pearls mean, then, not just to Sarah Siddons, Joshua Reynolds and visitors to the Royal Academy in the 1780s, but to pharaohs, Roman noblewomen, Mughal princes, Mongol conquerors and Hollywood royalty? What were they to the lustful and the bereaved, to the renegade and to the society matron? And where might Siddons's pearls have come from, in terms of their animal origins and their diaspora? The history of pearl fishing and trading is a global journey alighting in exotic locales: Tahiti, the Indian Ocean, the Caribbean, the Persian Gulf, the Red Sea. Over a century after the invention of cultured pearls, we might be blasé about such extravagant strands or mistake them for costume jewellery. However, in 1784 pearls could only have been natural. We realize with a jolt that this young woman is wearing a monarch's ransom around her neck. Her pearls might have been fished from the Persian Gulf, where prolific oyster beds sustained a pearl trade for 7,000 years. Maybe they were traded through the city of Zubarah, a tax-free port at the tip of Qatar that for a brief period in the late eighteenth century thrived on pearls. Or maybe the necklace was fashioned from much older stones, retrieved from the Caribbean, where almost three centuries earlier Christopher Columbus had searched for Oriental gold and found pearls instead. At first, the Spanish traded earthenware jars, scissors and beads for pearls; soon, they enslaved indigenous divers to retrieve them. Next, they imported Africans who were forced to dive without respite to ever greater depths. Perhaps Sarah Siddons's pearls were obtained by these divers who surfaced spewing blood.

Or maybe they came from closer to home. The British Isles had their own modest pearl fisheries. Scottish and Irish rivers and lakes yielded trout, salmon and eels, but also pearls, 'for lustre, magnitude and rotundity not inferior to oriental or any other in the world'.[12] In late summer, rural pearl-hunters would wade through shallow, clear water, prying mussels open with a sharpened stick. Findings were meagre, but occasionally an exceptional pearl might be discovered and sold at a country fair, traded onwards and bought by jewellers in London or Paris. As for the pearls' finders, they salvaged the mussel shells as spoons, while those born with silver spoons in their mouths – and those, like Sarah Siddons, who hauled themselves upwards – arrayed themselves in pearls.

BILLIE HOLIDAY

ASSOCIATED BOOKING CORP.
JOE GLASER, President
Squibb B'ldg 745 - 5th Ave. N.Y.   Beverly Hills, Cal.

Robin Carson, *Billie Holiday*, *c.* 1940, gelatin silver print.

Jean-Auguste-Dominique Ingres, *Joséphine-Éléonore-Marie-Pauline de Galard de Brassac de Béarn (1825–1860), Princesse de Broglie*, 1851–3, oil on canvas.

❀

ONE

The Oyster's Autobiography

Pearls are not the tears of mermaids or sharks, or princesses denied their lovers. They don't form in dragons' brains, or from lightning strikes on waves, or Venus shaking seafoam from her hair. They are not the love-children of oysters and dewdrops or fathered by moonbeams. For millennia, though, people speculated they might be.[1] The myth that a pearl grows when a grain of sand gets stuck in an oyster's soft body persists to this day. But oysters are filter feeders, siphoning plankton-rich water through their gills and adeptly ejecting sand. If grains of sand or the tears of every bullied princess made pearls, they'd be as plentiful as pebbles at the tideline. Far from being abundant, though, natural pearls are found in a scant fraction of oysters and other pearl-bearing molluscs. The chances of finding a large, unblemished, lustrous globe are vanishingly slim.

The way pearls *do* grow is just as magical as a dragon incubating a gem in its brain and as intimate as our own bodies. Like our bones and teeth, pearls form through biomineralization – the process by which living organisms create mineral matter. Perhaps this helps explain our long attraction to them. They are both miraculous and familiar. A diamond might have formed billions of years ago, hundreds of kilometres beneath our feet in unknowable heat and pressure. A pearl, though, is made by an animal that thrives in waters sometimes so shallow we can wade in them. A pearl can form in the

After Pierre Charles Trémolières (1703–1739), *Venus reclining upon
a fish in the sea; Cupid flies above her holding an oyster shell and pearls*, engraving.

time it takes a child to grow half a metre. The fire-flash of a diamond
recalls Earth's unvisited upper mantle. The gleam of a pearl recalls
an infant's milk tooth.

'The pearl is the oyster's autobiography,' claimed the Italian film
director Federico Fellini, defending his argument that all art is auto-
biographical.[2] We might think that an oyster's famously phlegmatic
and sedentary life lacks plot. ('You're enough to try the patience of
an oyster,' snaps a young crab to its mother in *Alice's Adventures
in Wonderland*.[3]) But thinking of a pearl as the story of an animal
responding to its ever-shifting environment might be the most helpful
way to understand its creation.

The oldest extant pearl, found in 2017 in the United Arab Emirates,
has been radiocarbon dated to between 5,800 and 5,600 BCE.[4] But
pearls' beginnings reach much further back to the Palaeozoic Era,

The Birth of a Pearl, 1923, stereograph.

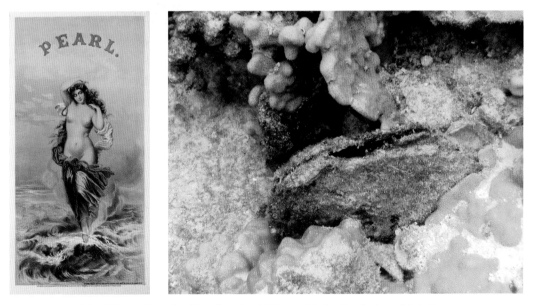

Major & Knapp Engraving, Manufacturing & Lithographic Co., *Pearl*, tobacco label, 1871.
Pearl oyster (*Pinctada margaritifera*), northwest Hawaiian Islands, 2004.

about 530 million years ago, with the emergence of molluscs. As popular on the menu today as through much of human history, this diverse group of soft-bodied invertebrates includes octopi, squid, snails, scallops, oysters and mussels. By the Cretaceous Period, 65 to 145 million years ago, pearls abound in the fossil record; the Age of Dinosaurs was also an age of pearls. Fossilized pearls from saltwater and freshwater molluscs have been discovered across the globe. They range from dull blisters on shell fragments to bumps slicked with iridescence and even to the rare glossy orb. Humans likely first encountered pearls as a tooth-cracking by-product of our earliest forages for food, and archaeologists still unearth them from prehistoric kitchen middens.[5]

There are more than 100,000 living species of mollusc and at least 60,000 fossil species, but very few bear pearls. Most molluscs capable of producing jewellery grade pearls are bivalves, and they owe this feat to their rudimentary but efficient anatomy. In essence, they are water filterers, and the unique qualities of water – its 'merroir' – build the character of the pearl. True to their name, bivalves have two hinged shells; their adductor muscles control the opening and closing of the shell valves. They strain water through their gills for food and respiration, and their soft bodies comprise organs for circulation, digestion and reproduction. Outside, the animal might be as prosaic as a pebble, but, as the French poet Francis Ponge marvelled, 'Inside a whole world awaits.'[6] The anatomy of this unassuming animal, arranged to maximize water circulation, leaves generous open space for the accidental formation of a pearl.[7]

A platter of glistening oysters-on-the-half-shell is likely to disappoint a pearl-hunter because edible oysters belong to the family Ostreidae, while pearl oysters belong to a different family, Pteriidae.[8] The genera *Pinctada* and *Pteria* bear the most valuable gems. Of these, a few superstar species birth the gems that have made their way to the ears of pirates and the necks of queens and Mughal princes. The length of a pinkie finger, the delicately rayed *Pinctada radiata*, or the

Fossil pearls collected in Florida, early Miocene to early Pleistocene epoch.

Cultured pearl in freshly opened oyster at a pearl farm in Hạ Long Bay, Vietnam, 2013.

ordinary pearl oyster, has pumped prodigal quantities of extraordinary pearls onto world markets.[9] Until the twentieth century, most saltwater pearls used for jewels and stores of wealth were from this species, which inhabits the Persian Gulf, home of the most fecund oyster beds, the Red Sea and the Indian Ocean. The diminutive *P. radiata* has played an outsize role in world history. When European colonists in the sixteenth century discovered vast banks of the pearls off the coasts of Panama, Colombia and Venezuela, they ignited a pearl rush fuelled by the transatlantic slave trade.

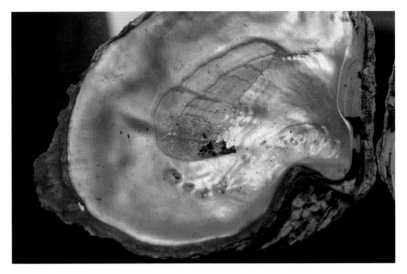

Pinctada maxima interior.

Pearls from *P. radiata* are typically under a centimetre in diameter. More bling South Sea pearls form in the dinner-plate-sized *Pinctada maxima*, the silver or gold-lipped oyster, which inhabits the eastern Indian Ocean and the tropical western Pacific. Black pearls come from *Pinctada margaritifera*, the side-plate-sized black-lipped oyster, which lives in some of the same waters. The La Paz oyster, *Pinctada mazatlanica*, found in the eastern Pacific between Mexico and Peru, produces silver-sheened pearls. *Pinctada fucata*, the small Akoya oyster, has been the workhorse of the modern cultured pearl industry.[10] These were the oysters that the Japanese entrepreneur Kōkichi Mikimoto used for the world's first industrial-scale pearl farming in the early twentieth century and still produce the canonical Japanese cultured pearl.

The maxim that irritated oysters make pearls is only partially correct, as not all pearls come from oysters. Abalone, a marine gastropod, can make tooth-shaped pearls as green-blue as spilt petrol. The queen conch, fished for its succulent meat, occasionally bears luscious, candy-pink pearls. The *Melo melo* sea snail, with its startling zebra-striped body curled in a yellow volute, can grow spherical flame-orange pearls, exotic as tropical fruits.

Pinctada Margaritifera, the black-lipped oyster.

Unknown artist, after James Bruce, *Pearls*, British, 1789, engraving.

Empty queen conch (*Strombus gigas*), Caribbean Sea, Bahamas, 2009.

Japanese Kasumiga freshwater cultured pearls.

We associate pearls with the sea, but freshwater mussels can grow pearls in shades of salmon pink, lavender, pewter and every nuance of white. For centuries, European aristocracy and the Russian Orthodox Church looked to freshwater mussels to help feed their colossal appetite for pearls. The European pearl mussel *Margaritifera margaritifera* was once endemic to cold, fast-flowing rivers from the Arctic to Portugal and from Russia to the northeastern seaboard of North America. Overfishing and industrialization led to its extinction in many rivers, and it is now classified as Critically Endangered in Europe.[11]

While saltwater Akoya pearls are still cultured in Japan and other countries, most cultured pearls are now farmed in Chinese freshwater mussels. The triangle pearl mussel, *Hyriopsis cumingii*, is a species as accommodating as rush-hour public transport, crowding more than thirty pearls into one mussel. China is the world's pearl-culturing behemoth, producing 3,540 tons, or more than 3 million kilograms, annually, representing over 98 per cent of global production. Of this, 99.5 per cent are cultured in freshwater mussels, nullifying the popular equation between oysters and pearls.[12]

The United States had its own fevered pearl rush in the mid-nineteenth century. The gems from freshwater mussels were of such high quality that they made their way onto European markets disguised as 'oriental' pearls from the Persian Gulf.[13] North America still has a dizzying variety of freshwater mussel species, though the ravages of the pearl rush and economic development have reduced the number of pearl-bearing species to around twenty, from a nineteenth-century high of three hundred.[14] Rare natural pearls can still be found along the Mississippi River basin, and have a niche following with artisan jewellers and their clients.

Whether freshwater or marine, all pearl-bearing molluscs grow pearls through biomineralization, the same process that forms their shells. Their genes determine this gemmological advantage over other

Chinese freshwater cultured pearls on display at a trade fair,
Costa Mesa, California, 2017.

molluscs. The creamy, briny flesh of an edible oyster is cradled by shell with the milky opacity of a bathroom fixture. But the inner shell of a pearl oyster shimmers like a rainbow. With the benefit of hindsight, it seems obvious that a pearl shares this nature, but it would take until the seventeenth century for scientists to affirm that a pearl and the inner layer of its shell are the same nacreous material.[15] Nacre consists of around 82–6 per cent calcium carbonate (the same material as eggshells), 10–14 per cent conchiolin (a protein 'glue'), 2–4 per cent water, and trace elements.[16] This material is secreted by cells in the animal's mantle – the thin membrane attached to the inside of the shell that envelops its soft inner body. When Plato wrote of 'the body, in which we are imprisoned like an oyster in its shell', he might have been thinking of the ruffed mantle, tightly binding flesh to shell.[17]

The classical world ascribed pearls to dewdrops seeping into the oyster, a myth that stubbornly persisted through the Middle Ages.[18] It was well into the nineteenth century before German scientist Theodor von Hessling insisted on the key role of the mantle.[19]

Nadine Leo, 14-carat gold ring with American natural
wing pearl and African orange garnet.

Jacques Callot, *Oyster on a Beach*, 1625–9, etching on paper, published by François Langlois.

Epithelial cells in the upper layer of the mantle secrete liquid that hardens to form the nacre that coats the shell's inner layer. When a tiny intruder, such as a parasite, gets lodged between the shell and the mantle, it can carry a few nacre-secreting epithelial cells further into the mantle tissue. These cells multiply around the foreign body as a form of defence, creating a cyst or 'pearl sac'. The sac begins its slow medicine, secreting layer upon layer of nacre around the intrusion, eventually forming a pearl. The pearl sac, not the parasite, makes the pearl. In fact, a pearl may develop in the absence of any foreign body, possibly triggered by abnormal growth of epithelial cells.[20] Cultured pearls – with rare exception, the only pearls now commercially available – can be grown without inserting anything other than tissue with these cells (see Chapter Three).

PEARL OF GREAT PRICE

Pearls can command eye-watering prices. It is the stuff of legend that the jeweller Pierre Cartier bought his flagship store at 653

Fifth Avenue, New York, with a double strand of his best pearls in 1917.[21] Christie's sold the storied pearl La Peregrina, once owned by Elizabeth Taylor, for nearly U.S.$12 million in 2011. Seven years later, a pearl and diamond pendant that had belonged to Marie Antoinette realized U.S.$36.2 million at auction.[22] Across cultures, a pearl is a useful metaphor for a rare and precious thing. The Quran promises that the faithful will be dressed in silk, gold and pearls in paradise.[23] The Gospel of Matthew compares the kingdom of heaven to a merchant seeking fine pearls: 'When he found one of great value, he went away and sold everything he had and bought it.'[24] Conversely, when jealous Othello murdered Desdemona, he 'threw a pearl away' (*Othello*, v.2). The Scottish poet Carol Ann Duffy distils desire into a servant girl zealously warming her mistress's pearls around her own throat before a party. The 'slow heat' of envy enters every pearl before she fastens the strand around her mistress's neck and retires to her attic bed:

> All night
> I feel their absence and I burn.[25]

Why do pearls incite such fervour? Why do we covet, crave and sometimes kill for something that shares the composition of old-school blackboard chalk or an antacid? Their rarity gets us part-way to an answer. When natural pearls were the only ones available, just one in 15,000 black-lipped oysters yielded a gem-quality pearl in French Polynesia. One in a million *P. radiata* in the Gulf of Mannar, off Sri Lanka, held an especially good pearl.[26] Since cultured pearls began saturating world markets in the twentieth century, the gem has become less rare, but it has lost none of its cachet. The journalist Stephen G. Bloom, who spent years tracking pearls' provenance across the globe, admits, 'I became pearl crazed, spending hours studying individual pearls till I got dizzy. Pearls were all I thought about. Wherever I went, I had vivid dreams about them.'[27] Just what about these gems, none of which had a flashy regal provenance, held his attention so long?

There are worldwide industry standards for grading diamonds based on clarity, colour, cut and carat weight. One might dream of a 10-carat diamond, internally flawless and colourless as rain. But a pearl's worth is harder to pin down. It is the only gem that emerges perfect from an animal, doesn't require skilful faceting and black ones can be the same price as white. You can peer into the cold soul of a diamond, but a pearl's beauty is as hard to judge as skin. To counter this imprecision, the gem industry has introduced systems to quantify a pearl's worth, such as the Gemological Institute of America's widely adopted '7 Pearl Value Factors' addressing size, shape, colour, nacre, lustre, surface and matching.[28]

LIGHT: MIRROR AND MERROIR

A pearl is a body of light, and its most arresting quality is its lustre. Mother-of-pearl, harvested from the inner layer of a mollusc's shell, has long been used as a shimmering decorative inlay. But a pearl's roundness accentuates its lustre. Light plays over the flat surface of mother-of-pearl, but a pearl glows from the inside. Poets have consistently drawn celestial comparisons. In Arabic, a particularly luminous pearl was termed *najm* – star.[29] Pearls were like the Moon and even thought to shine in harmony with its phases.[30] John Steinbeck evoked the unearthly quality of the gem's light when he described the 'pearly lucency' of dawn at Cannery Row in Monterey, California: 'it is the hour of the pearl – the interval between day and night when time stops and examines itself.'[31]

A pearl's light might seem supernatural, but its sheen derives from its natural structure and chemical composition. 'The pearl glows when the day is cloudy,' claimed Persian scientist and polymath al-Bīrūnī (973–c. 1052). But he also observed more prosaically: 'The fact is that the skin of the pearl is arranged layer upon layer, as in an onion.'[32] He was one of the earliest scholars to reconcile the mystical with the kitchen. Al-Bīrūnī provides great insight into the value the Islamic world placed on pearls, and his onion simile was on target, but today's powerful optical devices help us envision brickyards instead of chopping boards. A pearl's concentric layers are made of conchiolin and calcium carbonate in the mineral form of aragonite.

SEM (scanning electron microscope) image of typical nacre of an Akoya
pearl. The length of the scale bar is 5 μm (5 millionths of a metre).

An electron microscope reveals the thin sheets of hexagonal crystals
that comprise the aragonitic layers. These crystals line up like bricks,
bound with a mortar of conchiolin.[33] The animal is like a bricklayer,
each day completing one to seven courses until there are thousands.
Not all courses are complete, though: some stop midway; others
step higher, like ladders. Under magnification, the surface of a pearl
is more like a terraced hillside of rice paddies than a flat plain.While
this isn't visible to the naked eye, we can feel the gritty texture with
our teeth – the time-honoured way to tell a real pearl from a fake.

A pearl's optical qualities depend on this structure. Light reflecting off the pearl's uneven surface, like sunlight glancing off terraced
rice fields, contributes to its lustre. The reflections might be crisp
and mirror-like, with high contrast between areas of light and dark.
Or on a lesser-quality pearl, they might be soft and blurred, milky
or even chalky. A pearl's value is closely tied to its lustre, described
in consumer guides as 'analogous to brilliance in a transparent
gemstone.'[34] But a pearl has other optical properties to offer. Because
of the aragonite platelets' transparency, light can penetrate below

Pearl surface magnified ×50 showing characteristic whorl patterning.

the surface, bounce between the layers and split into the colours of the rainbow. This refraction of light creates the pearl's 'orient' or iridescence.[35] As al-Bīrūnī pointed out, 'the good attribute of a pearl is transparency so that the interior of it is reflected in the surface.'[36] Glow is far more than skin deep.

The animal depends on this architecture for more than good looks, though. A pearl is a relatively soft gem, measuring 3.5 to 4.5 on the Mohs scale, which ranks the hardness of minerals. A diamond, in contrast, measures 10, a ruby, 9, a topaz, 8. But despite its softness, nacre offers excellent protection to the animal it shelters. A predator can puncture or crack the outside of a mollusc's shell, but these cracks halt when they reach the interlocking 'brick wall' of nacre. The aragonite layers connect via mineral bridges through the organic 'mortar', further increasing the material's toughness and elasticity.[37] Inspired by nacre's tensile strength and fracture resistance 3,000 times that of aragonite, materials scientists are exploring biomedical applications.[38] Nacre is compatible with human bone and can stimulate new bone growth, making it a potential agent for bone repair.[39] Engineered materials could borrow its properties for bone replacements or other applications such as fracture-proof visors for space helmets.[40]

Under light microscopy, the surface of a pearl resembles a fingerprint whose whorls are the edges of the 'terraces' of nacre. All pearls, like fingerprints, are unique. Each has its pattern of nacreous layers created by an animal responding to its environment. Because cultured pearls are now so plentiful, we're tempted to think of them as commodities. But whether natural or cultured, all pearls are an expression of time and place, subject to natural phenomena beyond our control and miraculously singular. Merroir – the aquatic environment – is to pearls as terroir is to wine. Different species of mollusc make different types of pearls, just as different grapes make wine varietals. But water temperature, salinity, climate, weather, dissolved oxygen, the presence of nutrients, minerals and pollution all craft the pearl, just like soil, topography and sunlight craft the wine.

We might find cold water bracing, but it turns an oyster sluggish.[41] A lowered metabolism, however, is excellent for nacre quality. In warmer temperatures, in which oysters can more readily take up calcium carbonate from the water, nacre is laid down faster, making a bigger pearl.[42] But size isn't everything. When nacre accumulates too quickly, imperfections give the surface a hammer-beaten appearance. Cold water decreases the nacre tablets' thickness and growth rate, resulting in a more uniform surface with radiant lustre and iridescence.[43] As the last few layers of nacre can perfect the pearl, many farmers harvest in winter. In Japan, cultured Akoyas are typically harvested between November and February, with the famous *hamage* auctions beginning in December. It isn't just pearl farmers who take advantage of the correlation between nacre and water temperature. Scientists have turned Jurassic fossil shells into paleothermometers, reading Earth's temperature millions of years ago from the formation of the shells' nacreous layers.[44]

Just as certain regions are renowned for their wines, places became famous for the lustre of their pearls. A pearl fishery existed in Bahrain, an archipelago in the Persian Gulf, as far back as the first millennium BCE. The islands became synonymous with the world's finest pearls. The seventeenth-century French gem merchant Jean-Baptiste Tavernier, whose extensive travels took him to the world's jewel entrepôts, was among the many travellers, merchants and jewellers who claimed Bahrain pearls were unrivalled. Their prestige owed in part to merroir. In Arabic, Bahrain means 'two seas', and it may be the site mentioned in the Quran where pearls and coral marked the confluence of two bodies of water.[45] Around Bahrain, submarine freshwater springs bubble up to meet salt water, a phenomenon that turned the islands into an Edenic garden bursting with figs, dates, citron, almonds and peaches.[46] Bahrain's miraculous waters were considered sacred and its pearls a blessing, but a natural explanation was that the mingling of waters maintained just the right salinity levels (25 per cent–27 per cent) we now understand as crucial for perfect lustre.[47] 'Of the first water' (of the highest quality) is an ancient figure of speech that originally referred to jewels. When used of pearls – grown in water by an animal whose body is 90 per cent water – the idiom is perfectly accurate.

A pearl is an autobiography that records an animal's waters of origin, its feasts and famines, successes and tribulations. Temperature, water currents, sunlight, plankton and minerals all feed into its narrative arc. Even the ancient association with the Moon may have somephysiological truth: recent research suggests that oysters synchronize behaviour according to lunar cycles and are sensitive not just to the Moon's light but its phases.[48] Exquisitely tuned to its environment, a mollusc is particularly susceptible to environmental degradation. Algae, disease and pollution can kill molluscs, and there are records of mass mortalities affecting millions of animals, notably in Japan during the 1990s. There are also well-documented examples of the effects of pollution on nacre production. In March 1970 the oil tanker *Oceanic Grandeur* ran aground in the Torres Strait off the north coast of Australia. More than 1,100 tonnes of oil spilt into the sea. Most of the oysters died, and the survivors developed shell and nacre deformities.[49]

ROUND OR CROOKED-WAISTED: SHAPE AND SIZE

All pearl-loving cultures understand that each pearl has its distinct personality. The naturalist Pliny tells us that the ancient Romans used the term *unio*, the unique gem. Al-Bīrūnī knew over fifty different terms for pearls in Arabic and Persian. They ranged from *nutfah* (drop), to *pashki* (olive-shaped), *mudarras* (like a molar tooth), *shahwar* (regal) and *'uyun* (pearls as round and glistening as the pupils of eyes).[50] There are pearls so charismatic that they have celebrated names, such as the Hope Pearl, a silver-tipped mammoth named after its collector Henry Philip Hope; La Peregrina (the Pilgrim) owned by a succession of celebrities; and al-Yatīma (the Orphan), an unmatched beauty renowned in the medieval Islamic world.

If we were to believe the evidence of hundreds of years of portraiture, we would think that all pearls were large, white and round, or shaped like prize Anjou pears. But the pearls we see around the necks of Sarah Siddons and aristocrats down the ages were outliers. Such pearls would have been the cream of the crop, sieved from great quantities of smaller gems of different shapes and undoubtedly enhanced by the artist to amplify the sitter's status. The existence of words like *lawzi* (with slender ends), *kamarbast* (crooked waisted) and *jawdanah* (grain of barley) betrays the fact that few pearls were perfect orbs.

The perfectly round, though, have commanded top prices. George Frederick Kunz (1856–1932), renowned mineralogist and Tiffany's Vice President of Gemology, noted that a top-quality pearl should be 'as even in form as though it were turned on a lathe'.[51] But like al-Bīrūnī, Kunz conceded these were rare and listed colourful descriptive names for many non-spherical shapes, such as haystacks, wing pearls, feather-shaped, turtlebacks and strawberries.[52] Kunz was writing in the early twentieth century when about one in a thousand natural marine pearls was round. Since that time, expectations have grown for cultured pearls to achieve this ideal sphericity, described in the industry as 'eight-way rolling', reinforcing our belief that pearls are round.[53] Culturing pearls artificially kick-starts the biomineralization

Feather freshwater cultured pearls.

process. For saltwater pearls, a round bead is implanted with a fragment of mantle tissue from a donor oyster, increasing the odds of roundness to one in ten cultured pearls.[54] These round pearls are the ones that reach the market and inform our ideas of value and norm.[55]

Natural pearls are more apt to go rogue. They can squeeze between the shell and mantle to form a blister pearl. Sometimes nacre entombs entire fish against the shell.[56] Kunz observed that when pearls grow near the hinge, they are elongated like dogs' teeth or birds' wings.[57] Mississippi Valley freshwater mussels frequently grew pearls of this shape, which Kunz attributed to excessive calcium carbonate in the river.

If they form close to a muscle, they are likely to become misshapen baroques. 'Baroque' now refers to a European art movement of circa 1600 to 1750, but the word emerged in the sixteenth century in the Spanish-controlled Caribbean pearl fisheries, where for tax purposes *barrueca* referred to an irregular pearl of lower worth than a round one.[58] Renaissance jewellers made a virtue of deformity, creating golems from clumps of nacre. Honouring their aquatic origins, they became dolphins, mermaids, sea monsters or swans. They

Pendant in the form of a swan, northern European,
16th century, gold, partly enamelled, pearls.

morphed into parrots, cats, the crucified Christ or the Holy Spirit. Orbs were obvious, but it took a jeweller's ingenuity and a collector's taste to discern the puffed down of a swan in a pearl or the manacled body of one of the enslaved African divers on which the Spanish fisheries depended.

The shape of a pearl is a connection back to the animal in whose body it grew. A baroque pearl embodies the passage of time required to coat an intrusion or fuse two or three pearls into one. Tahitian black pearls often have pronounced circular grooves, inherited from their years in the pearl sac, slowly rotating as they formed, like and wholly unlike a human child.[59] Even when deviating from the desirable round, their shapes are never accidental but part of the pact between the animal and its environment. Fashion sways how we value this. The current trend for Tahitian baroque pearls, for instance, is shaped by the ongoing shift towards informality in dress, admiration for the 'natural' and the region's status as a popular tourist haven.

Our ideas about the size of pearls are as skewed as our expectation of their roundness. Aquaculture and the selective breeding of bivalves have redefined the norm, bringing us larger cultured pearls. We're used to seeing photographs of celebrities or politicians in unmissable South Sea globes. Our understanding of the scale of jewellery is often distorted, as blown-up visual images of such intimate objects tend to inflate their size. Natural pearls are surprisingly small. Many were less than 3 millimetres in diameter, and anything over 8 millimetres was rare.[60] The smallest used in jewellery were seed pearls. Kunz noted that these were so tiny that 18,000 might weigh just 1 ounce (28 g) and that they were shipped in hanks of a dozen strings.[61]

Brooch, German, *c.* 1700–1720, silver, enamel, lacquer, pearl, ruby or simulant.

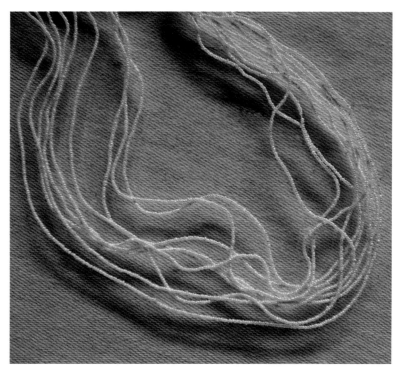

Seed pearl strands comprised of pearls only 0.7 mm in diameter.

Even smaller were dust pearls, which, however perfect in form, were only useful for medicines. These specks could weigh less than 1/25 of a grain. The jeweller's grain – 50 milligrams – was the traditional weight measurement for natural pearls and possibly referred to an ancient standard for the weight of a grain of cereal.[62] Four grains make a carat, the glitzier unit of weight for diamonds. Seed pearls might weigh half a grain and were sold in lots. Anything over four grains was sold individually. Kunz noted that perfect orbs weighing one hundred grains were termed 'paragons' and could reach three hundred grains or more, though usually at the cost of lustre.[63] The exceptional La Peregrina, the size of a pigeon's egg, weighed in at 202 grains. The largest pearl ever recorded, the brain-shaped Pearl of Lao Tzu, was recovered from a giant clam in the Philippines in 1934 and reached a staggering 127,574 grains. However, the Gemological Institute of America sniffily terms it a 'chalky concretion'.[64]

Cultured pearls sold in bulk are weighed in momme, a traditional Japanese unit retained by the pearl trade. Seventy-five grains make one momme. Japanese factories purchase pearls by the unglamorous plastic bagful from farmers or agents, drill, sort and string them, then offer them for sale by the momme. In the high-volume world of wholesale pearls, 1,000 momme make up one kan.

COLOUR

For much of history, the ideal pearl has been large, round, glossy as a cherry and white – possibly with a hint of gold. From European royalty to Mughal emperors to Holly Golightly with a pastry outside Tiffany's, white pearls were the desirable norm. Their colour carried a raft of socially fabricated meanings: innocence, purity, poise, perfection, cleanliness and its mate godliness. Poets, portraitists and historians crystallized the connection between the white gem and virtue. In the Middle English poem *Pearl*, a beloved deceased daughter appears in a dream to her grieving father. Dressed in pearls as a bride of Christ, she is an apparition of unblemished white.

The modernist Argentine poet Alfonsina Storni revisited the connection between pearls, chastity and whiteness in her poem 'You Want Me White' ('Tú Me Quieres Blanca'):

> You want me dawn,
> Want me foam,
> Want me pearl.
> To be a white lily
> Above all, chaste.
> Of tenuous perfume.
> Corolla closed.[65]

The *Pelican Portrait*, of circa 1575, an overt piece of painted propaganda attributed to the English court portraitist Nicholas Hilliard, captures Queen Elizabeth I in the guise of the Virgin Queen married to the state. From her chalky make-up to the pearls studding her

Unknown artist, pendant representing the Pelican in Her Piety, Spain,
c. 1550–75, enamelled gold, set with a ruby simulant and hung with pearls.

dress and draped around her shoulders, to the sacrificial pelican
(a symbol of Christ) pecking blood from her breast to feed her brood,
the portrait is a fetish of whiteness. Al-Bīrūnī, too, connects white
pearls and divinity, citing theologians who explained that Allah had
made pearls like the white of an eye.[66]

From encyclopaedists to chemists, experts were fairly clear that
pearls were white-ish. The Italian astronomer Camillo Leonardi,
writing in the early sixteenth century, insisted that the pearl 'has
the first place among white gems' and that the best were 'white like
polished silver'.[67] He dismissed British pearls as 'dull with a certain
whiteness'. A pale neutral 'pearl' was a popular choice of paint for
eighteenth-century interiors.[68] In the nineteenth century, George
Field's classic treatise on pigments *Chromatography* (1835) listed
'pearl white'. A 1906 manual for painters included a recipe for 'pearl

grey' – forty parts lead white, five parts vermillion, one part chrome green.[69] For American colour impresario Albert Henry Munsell, whose famous Munsell system aimed to standardize colour internationally, 'pearl' was classified under grey.[70] The entry for 'pearl' in Ian Paterson's authoritative twenty-first-century *Dictionary of Colour* is 'light greyish blue'.[71] Pantone, the gold standard in universal colour matching, lists various pearled hues, but pearl itself is a delicate blush pink.[72]

While a pearl can be any shade of white, pastel pink or grey, it can also be gold, silver, black, aubergine, petrol blue or peacock green. Writing when American freshwater pearls were still plentiful, Kunz saw every hue and shade, a chromatic bounty he thought could only be preserved by a poet. Today, visiting a major gem show or leafing through an industry publication confirms an abundance of white pearls – but also a sweet shop of pistachio, candyfloss, lemon, chocolate and lavender ones.

A pearl's colour comes from a complex mix of factors, but the most important is the species of mollusc that produced it. Pigments in the conchiolin layers are responsible for colour (aragonite platelets are typically transparent), and these are genetically determined.[73] The colour of a natural pearl will usually match the inner layer of the animal's shell as the same cells are responsible for secreting nacre on both surfaces. Cultured pearls, though, involve two animals – the donor of the mantle tissue and the living mollusc that will grow the pearl. Through careful selection of the all-important mantle donor, farmers can manipulate the colour of the finished gem. Merroir comes into play, too, as trace elements such as iron and magnesium in the water can alter the hue.

Colour sways a pearl's value, and value is always culturally constructed, as the giddy history of gem trading makes as clear as an aragonite crystal. The crisp whiteness we have come to expect in cultured pearls is, to an extent, engineered to suit Western tastes. Japan's Akoya oysters tended to produce creamy pearls, but the cultured pearl industry, orientated towards European and American consumer markets, encouraged the pursuit of a cooler alabaster.[74]

A 1936 advertisement for Bergdorf Goodman, the New York luxury department store and arbiter of taste, touted its stock of Mikimoto pearls 'shaped like the full moon and whiter than the morning star', priced from $28 to $5,000 – an oddly profane use of the Christ/white/pearl association.[75]

Developments in periculture, or pearl farming, can improve the yield of a desirable colour. Japanese farmers learned that whiter pearls were harvested in enclosed bays rather than from waters with strong ocean currents, and they noticed a correlation with temperature fluctuations, especially cold winters.[76] (Millennia earlier, Pliny had argued that brilliant whiteness depended on cold, clear water.[77]) The tranquil inlets of the Shima Peninsula's Ago Bay on Honshu island, Japan, have been favoured culturing sites since the early twentieth century, but oyster breeders have gone further in their pursuit of white. Wild Akoyas are typically brownish-purple with spots or stripes. Rarer white shells are caused by albinism, which is a genetically inherited trait. Through selective breeding, farmers can boost their harvest of wintry gems.[78] Later, during processing, technicians remove blemishes or unwelcome colour casts through chemical treatments such as bleaching with hydrogen peroxide or ultraviolet light.[79] As with dentistry, though, it takes skill to know when to stop the process. Overly bleached teeth and pearls look unnatural, but with pearls, there is the added risk of permanently damaging the conchiolin.

Taste is fluid, and not everyone admires optic white. Around the Persian Gulf, people preferred pearls with a slightly golden tint, believing that cold white looked too much like stone.[80] In mid-nineteenth-century Bombay, the world's major pearl-trading market, perfect spheres with a yellowish hue were the most prized.[81] In the age of cultured pearls, the same species can be manipulated to acquire different hues and appeal to other markets. Akoyas can be bridal white, sensuous gold or sophisticated silver. South Sea pearls can be lunar white, but the same species, *P. maxima*, can produce honeyed gold orbs.

White is always in season. But sometimes pink is too. One of the most flamboyant and rare pearls comes from a marine gastropod

Tiffany & Company, sautoir with pearl pendant,
c. 1900–1910, queen conch pearl, platinum, diamonds.

Detail of Tiffany & Company sautoir with pearl pendant.

native to the Caribbean, *Strombus gigas*, the queen conch. Its cycles of cachet and neglect are a reminder that while pearls have always been treasured, colour is more subject to fashion. European colonists discovered the sensual, pink-lipped shells in the Caribbean, where islanders appreciated conch as a culinary delicacy. Soon these symbols of the lush tropics entered connoisseurs' collections, appearing in Dutch landscape paintings, English cabinets of curiosities and Italian grottoes and fountains. These large snails congregated in colonies of several hundred on seagrass and, as late as the 1970s, could be scooped from the warm shallows. But the odds of finding even a gritty 2-millimetre conch pearl are one in a thousand. One in 20,000 animals grow larger ones, which can reach a gumball-sized 20 millimetres. The most admired are punch pink with a shimmering surface like moiré silk.

Christopher Columbus probably meant this gem when he referred to '*la perla rosa*', but their scarcity saved their makers from

Sarah Ho, 'Candy Conch Pearl Ring', conch pearl set between four diamonds in an 18-carat rose gold ring with white enamel bands.

a *P. radiata*-style ransacking.[82] The first recorded Western collector
was uber connoisseur Henry Philip Hope, of Hope Pearl and Hope
Diamond fame. Several conch pearls, ranging in colour from yellowish (probably produced by juveniles) to fine pinks, were listed in the
1839 catalogue of his collection.[83] Conch shell had been fashioned into
jewellery since the Renaissance, and carved pink 'rosaline' cameos
were popular in the Victorian era, but the rare pearls received a boost
when celebrity influencers adopted them. Pearl aficionada Queen
Alexandra, wife of Edward VII, was particularly fond of conch pearls.
George Frederick Kunz showcased the rosy specimens at the Chicago
World's Fair in 1893, and for a couple of decades they were all the rage
with art nouveau jewellers before slumping into disfavour.[84]

In the 1980s they found an unexpected champion in the palaeon-
tologist Susan Hendrickson, better known for discovering the world's
most complete Tyrannosaurus rex fossil, now in the Field Museum
in Chicago and named Sue in her honour. In addition to unearthing
fearsome carnivores, she revived the taste for a placid herbivore's pink
jewels. Elizabeth Taylor cemented their reputation when she mod-
elled earrings and a necklace by celebrity jeweller Harry Winston on
the September 1990 cover of *Ladies' Home Journal*. Taylor smoulders
against a backdrop of Barbie pink, the conch gems as rosy as her lip
gloss. Since then, Mikimoto has offered a very exclusive range of
jewellery, pairing warm-hued conch pearls with diamonds and cool
platinum or white gold.

Black pearls have experienced even greater swings in fortune.
A black pearl, like a black swan, was a puzzle to Europeans. The
globe-trotting gem merchant Jean-Baptiste Tavernier thought they
grew from oysters anchored in dark mud, though he was disappointed
that Europeans seemed indifferent to them.[85] Others believed they
resulted from an animal's illness or advanced age or even from squid
ink contaminating the water.[86] Although they could indeed be jet
black, they could also shimmer between aubergine, peacock green
and petrol blue. They derived their colour not from dirt or disease
but because *Pinctada margaritifera*, the black-lipped pearl oyster,
secretes dazzling dark nacre. This oyster inhabits the Red Sea and

the Persian Gulf but was never exploited like *P. radiata* in these seas. Black pearls did, though, have a refined cult following. In his guise as Tiffany's august gemmologist, Kunz had unparalleled access to the world's most exclusive collections. In 1891 he examined the Russian crown jewels and deemed Catherine the Great's collection of black pearls 'unrivalled'.[87]

In the South Pacific, around the areas that are now French Polynesia and the Cook Islands, the black-lipped pearl oyster is prolific. Spanish and Portuguese navigators visited these islands in the sixteenth century. The British claimed Tahiti in 1767, the French a year later. Europeans questing for tropical commodities and missionaries questing for souls noticed islanders wearing the curious black gems, but there was little overseas demand until Tahiti became a French protectorate in 1842 and black pearls reached French markets. Rarer than white pearls, they caught the fancy of Empress Eugénie of France, wife of Napoleon III, and an exclusive trend was born. They were the photo negatives of white pearls – chic, mysterious and as exotic as the atolls France was steadily adding to its territories.

European entrepreneurs stripped black-lipped oysters from the lagoons not just for pearls but for the dusky mother-of-pearl that fascinated European consumers. Shipped out by the ton, mother-of-pearl was used for buttons and inlay. Proceeds from pearling lined the colonial administration's coffers and European traders' pockets, but the Polynesian islanders weren't getting rich. Divers were caught in a vicious cycle of servitude, as merchants extended them credit then paid them a pittance for their haul.[88]

In response to the depletion of oyster beds, the French government encouraged oyster farming and pearl culturing experiments. But while Mikimoto's dainty white pearls found large middle-class markets in Europe and the United States, the black pearls cultured by French entrepreneurs on farms in French Polynesia struggled to find buyers. People were suspicious of their origins, and no jeweller would promote them.

This changed when marketing wizardry put the emphasis firmly back on the myth of unsullied Polynesian nature. In 1973 the

Italian-born godfather of the pearl trade, Salvador Assael, offered cultured black pearls to New York's most blue-blooded jeweller, Harry Winston.[89] Initially reluctant, Winston bought eighteen strands. Assael, who had his own stake in Polynesian pearl farming, wove the artful seduction that black pearls were from Tahiti's turquoise waters. In fact, they were farmed in remote lagoons, but Tahiti was a tourist haven, and 'French Polynesia' wrapped Gallic chic in tropical charm.

Van Cleef & Arpels, Cartier and Tiffany picked up the scent, and soon New York fashionistas were paying hundreds of thousands

Tahitian cultured pearls.

of dollars for the latest must-have dark gem. These were, of course, cultured black pearls. But Fifth Avenue celebrity jewellers presented them as the almost priceless results of a partnership with nature. Advertisements announced, 'A New Gem is Born'.

Convincing a sceptical public that these dark orbs were as desirable as white ones was a marketing triumph. They were adopted by luminaries such as Isabella Rossellini and Elizabeth Taylor (again), who in 1996 launched her Black Pearls perfume: 'rare, sought after, coveted, from a world apart'. This was certainly true of the pearls. In the mid-1980s, the best-quality black pearls were double the price of the most expensive white South Sea ones. By the end of the millennium, annual production was worth $170 million, and pearls cultured from *P. margaritifera* became French Polynesia's highest source of income after tourism.

THE PROMISE OF SUSTAINABLE PEARLS

But it wasn't only the pearls that were expensive. The harsh environmental and social costs of this rapid growth soon hobbled the industry and the region's new wealth. The ensuing collapse and partial recovery of pearl culturing in French Polynesia is a case study in the long-term rewards of adopting more sustainable practices that acknowledge that pearls, like coral, are renewable – not limitless. Before the invention of culturing, people for millennia either decimated natural pearl beds or fished with restraint, respecting the stocks' ecological limits. They conserved resources by choice, allowing juvenile molluscs time to mature, or they were coerced by such horrors as the infamous pearl gallows of seventeenth-century Bavaria.[90] These gibbets were erected beside rivers controlled by the royal pearl fisheries as a deterrent, their locations noted, with a shiver, by writers of guidebooks to the bucolic pleasures of the region.[91]

The shift to pearl culturing in the twentieth century brought its own set of problems: disease-susceptible monocultures, over-supply and deflated prices, and the ongoing challenges of degraded habitat. Faced with a toxic combination of some of these challenges, the industry in French Polynesia was forced to reform. Between 1986 and 1994, small farms had multiplied tenfold.[92] In 2000, at the height of the boom, the industry employed over 7,000 people and produced 11,364 kilograms (25,054 lb) of pearls, twenty times the amount for 1990.[93] But a glut of small farms crowded the market with poor quality, thin-nacred stones, badly tarnishing the industry's reputation. Inspired, perhaps, by Kōkichi Mikimoto's famous incineration of inferior pearls in Kobe in 1932, the Polynesian government staged the dumping of 35,000 pearls into the ocean in 2000.[94] This stunt did nothing to staunch the decline, and within a decade, prices haemorrhaged by 75 per cent.

Many pearl farmers lost their livelihoods, but the upheaval created the impetus for change and a healthier future. Recognizing the need to differentiate their product, a small number of farmers have adopted a different approach, aligning themselves with rising eco-awareness. By common estimate, 95 per cent of a farm's income comes from just 2 per cent of its pearls, so these farmers work with the delicate harmony between healthy oysters and ocean biota to beat those odds. Oysters suspended in baskets are susceptible to biofouling – plants and animals that bore through shell, clog the oysters' valves,

Kamoka Pearl Farm, Ahe, French Polynesia.

reduce water and oxygen flow, and break ropes.[95] The solution at industrial farms is to blast the shells with pressure hoses, damaging surrounding environments.

But pearl farms in French Polynesia are located on remote atolls whose reefs team with fish. Eco-friendly pearl farmers take advantage of reef fishes' free labour, allowing them to nibble away pesky accretions. This symbiosis encourages reef fish to thrive, and consumers find the narrative of pearl farms and healthy reefs a compelling reason to pay a premium for pearls grown in these waters.[96]

Conscientious growers reduce their carbon footprint by harnessing solar and wind power and using rainwater to supply their freshwater needs.[97] Instead of shipping nucleus beads from North America and further endangering American freshwater mussels in the Mississippi River basin, these farms use mother-of-pearl recycled from local oysters. They outperform all other nuclei by a factor of three.[98] Local communities benefit from the boost to their economies. Before pearl culturing, many young people migrated to Papeete, the capital of French Polynesia, in Tahiti to seek work, but well-paid employment at pearl farms has allowed many to return to the atolls and their families.

Grafting work at the Kamoka Pearl Farm in Ahe, French Polynesia.

Underwater at the Kamoka Pearl Farm, Ahe, French Polynesia.

For these farmers, sustainability means taking the long view and operating with respect for fragile ecosystems. They craft a compelling message about pearls' role in improving the natural environment and the lives of each person through whose hands the gems pass. They secure a premium for their product, help their employees develop new skills and steward natural resources. They educate consumers to spend discriminately and mindfully, acknowledging every resource as precious. Sustainable pearls shift the narrative away from commodity, and firmly back to the autobiographies of oysters and oceans and the people who tend them. The moment pearls are prised from their shells, the stories of their makers turn to their takers.

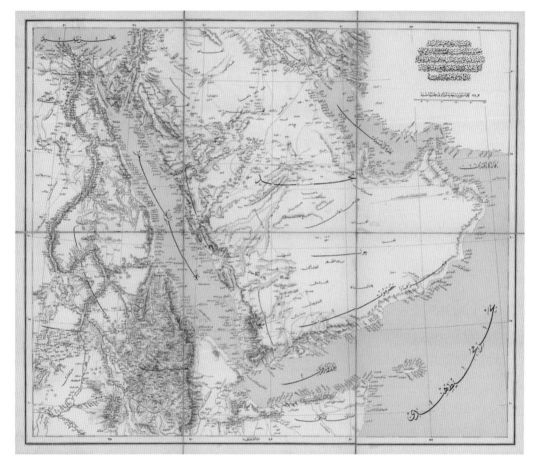

Erkan-I Harbiye-I Umumiye (General Staff of the Ottoman Army),
Arabian Peninsula and the Red Sea (1897).

TWO

❦

Harvest

In the early afternoon of 26 May 1932, a violent protest erupted at Manama, the modern capital of the island nation of Bahrain.[1] The previous day, a desperate group of pearl divers had confronted Sheikh Abdullah, a member of Bahrain's ruling family, complaining of poor wages and oppressive working conditions that left them permanently indebted to their employers. They received short shrift, and the ringleaders were detained at Manama's police station. The following day, a mob of 1,500 disgruntled men armed with clubs and iron bars smashed their way into the station. They rescued one of their leaders and tried to break into a storeroom containing rifles and ammunition. News accounts later reported that the disciplined local police avoided mass bloodshed by refraining from shooting into the crowd. However, in the melee, trigger-happy security guards killed two divers and injured at least five others.

The rioters were beaten back to the shore and fled by boat. On their way back home to the island of Muharraq, they were intercepted by Sheikh Abdullah and ordered to turn around. Jeering, the men lifted their robes and waggled their penises at the affronted prince (the universally recognized gesture for Go F*ck Yourself). Once they reached shore, most melted into the crowds. For a brief period, the town rang with gunfire and police chased protestors through Muharraq's narrow, twisting alleyways. They arrested several fugitives and quickly restored order. Sheikh Abdullah's brother, smarting from the blow to his family's honour, ordered a rebel safe house burned to the ground and the immediate arrest of anyone challenging their terms of employment.

The rebels were dismissed as 'ruffians and semi savages from a dozen different countries', and the insurgency was blamed on weak government.[2] When the diving season opened in early June as it had for hundreds of years, this troubling chapter seemed closed. Except, within a few years, the pearl industry – the region's economic engine – stalled. By the 1960s, it was over. The surprise was not that the divers rebelled in 1932 but that they suffered so abjectly for so long. Through most of human history, the 990-kilometre-long (615 mi.) Persian Gulf had been the source of most of the world's pearls, and the trade had earned the region fabulous wealth. The boom years weren't centuries in the past, but within living memory in the early twentieth century. The oyster beds were so extravagantly productive that in Kuwait, 1912 was known as the Year of Tufha – the year of superabundance. That year, a single pearl in Paris sold for $175,000.[3] Its value, though, had snowballed the further it travelled from its waters of origin. It is a universal law that the hands that harvest – whether pearls, coffee or asparagus – don't belong to the bodies getting rich. Those who extract gems from the sea, stones from the earth or food from the soil rarely reap what they deserve for their labour.

DEMAND AND DEBT IN THE PERSIAN GULF

The history of pearl fishing in the Persian Gulf stretches back at least 7,000 years.[4] In Qurum, in northern Oman, archaeologists have ex-cavated burial sites dating from the fourth millennium BCE. At one cemetery, the skeletons lay on their sides facing the sea, covered in the bones of fishes and green turtles – perhaps a funerary meal. Some held a single pierced pearl clenched in their right fist – an ancient pact between human and ocean.[5]

Many ancient myths claim that pearls are magical and confer imm-ortality, none more evocative than the 4,000-year-old Mesopotamian odyssey the *Epic of Gilgamesh.* This poem was composed over a time span when luxury goods, including gold, ivory, precious stones, dates, pearls and aromatic plants, were traded through Bahrain to the ancient cities of southern Mesopotamia.[6] Towards the end of the poem, an exhausted Gilgamesh learns that a prickly plant growing on the seabed holds the secret to rejuvenation. On hearing this,

> He tied heavy stones [to his feet],
> They pulled him down into the watery depths
> He took the plant though it pricked [his hand].
> He cut the heavy stones [from his feet],
> The sea cast him up on his home shore.[7]

Early accounts of pearl diving in the Persian Gulf mention the use of such weights, and for a text delighting in wordplay, the plant that lacerates hands is likely an oyster cradling the cure to mortality.[8]

During the conquests of Alexander, Persian Gulf pearls were lucrative booty.[9] Around this time, pearls were also fished off Indian and Sri Lankan shores in the Gulf of Mannar, another location in the Indian Ocean associated with pearls for centuries. Still, this region rarely challenged the Persian Gulf's status as the world's purveyor of pearls. By the Roman era, rivers of pearls flowed westwards from both gulfs to meet the empire's insatiable appetite – to the dismay of moralists like Pliny, who considered them wasteful things fit only for women.[10] Pearls became a global commodity, feeding markets throughout the Middle East, India, China and the Mediterranean. On the bodies of Persian princes, Indian aristocrats and Roman society matrons they were status symbols – for which the bodies of motley 'semi savages' paid dearly.[11]

From the days of Gilgamesh descending with stone weights, Persian Gulf pearl-diving traditions and techniques were relatively unchanging. The annual diving season opened between mid-May and early June and continued while the waters remained warm until September or October.[12] Divers, many of whom were seasonal migrants from throughout the region, joined the crews of boats which

Jan van der Straet, called Stradanus, *Diving for Pearls in the Persian Gulf*,
1594–6, pen and brown ink, brush and wash on paper.

departed, sometimes in their thousands, to the banks of *P. radiata*
(and more rarely, *P. margaritifera*) that flourished throughout the
Gulf. Nineteenth- and early twentieth-century maps chart hundreds
of oyster bed locations fringing both Persian and Arabian coastlines,
as well as far into open waters.[13] They clump around Bahrain and the
'mitt' of Qatar and cluster around the rocky tip of Oman, where it
juts into the Strait of Hormuz. The most favourable sites were level
stretches of fine white sand on top of coral. The oysters anchored
themselves to this substrate or clung together in a mass called a *tabra*;
one lucky *tabra* could salvage an entire poor season.[14]

In the nineteenth century, large vessels up to 30 metres (100 ft)
long were crewed by around 65 men. Almost half were divers, about
twenty were rope-haulers and the rest were deckhands, cooks and
a captain (*nakhuda*).[15] The men were local Arabs, Persians, Indians
and Bedouin nomads who travelled to the coasts for the season. The
crews included enslaved Africans, essential labour for an industry
that during its boom years in the early twentieth century employed

around 50 per cent of the male population of coastal areas.[16] By 1929 roughly 20,000 enslaved divers, many from East Africa, worked the oyster banks each season.[17] Some had been kidnapped as young boys and forced to dive from their early teens; some were born into slavery. As they had to surrender their earnings, even the most successful found it difficult to earn their freedom.

Onboard ship, the working day began after early morning prayer. The crew's first noxious, fly-ridden task was to prise open with a curved knife the oysters caught the previous day, extract any pearls and tip the shells overboard to 'nourish' the beds. After a light breakfast of dates and coffee, the divers began work outfitted with the same equipment gulf divers had used for centuries – a horn nose clip, leather finger guards, a basket and a knife. They plugged their ears with oil or beeswax-soaked cotton and descended feet-first, aided by a rope with a stone weight, which a hauler brought back up once the diver reached the bottom. For the 40 to 75 seconds they could hold their breath, they gathered as many oysters as possible (generally between three and twenty) and then jerked on a second rope to return to the surface. They repeated this routine up to fifty times, resting a little between plunges, breaking the day with coffee and an occasional cigarette.[18] The crew worked, ate, slept and prayed aboard ship, shared cramped quarters with cockroaches, endured the reek of dying shellfish and blistered under the hot sun. Alan Villiers, an author who spent a year sailing with Arab crews just before the outbreak of the Second World War, likened the deck of a Kuwaiti pearler to a New York subway platform at rush hour with no trains arriving.[19]

The season concluded with the fleet racing home in autumn. In some ports, it was traditional for the boats to gather outside the harbour and enter together, welcomed home by jubilant crowds and music. After a rest, some crews then sailed south, extending the season in the warmer waters of the Gulf of Mannar, between Sri Lanka and India's southeastern tip.[20] Divers had gathered pearls off India and Sri Lanka as far back as the fourth century BCE.[21] The same *P. radiata* pearl oysters flourished in these waters, and communities practised

Alfred William Amandus Plâté and Co., pearl divers,
c. 1890, from *The Hundred Best Views of Ceylon.*

many of the same diving techniques and traditions as in the Persian Gulf. There were important differences, though, in the industry's organization. In the Persian Gulf, the oyster beds weren't the property of any person or state and were theoretically open for anyone to fish. Small and medium-sized businesses operated with little government interference. In contrast, pearling in the Gulf of Mannar was strictly regulated, which had consequences for the retrieval of pearls. Unlike their Persian Gulf counterparts, fleets working off Sri Lanka returned to port each day with their oysters unopened, allowing government regulators oversight of the haul. Many visitors to the Gulf of Mannar's pearl fisheries complained of the stench of the piled oysters rotting on the beaches. The Portuguese officer João Ribeiro, who spent over a decade in the region in the mid-seventeenth century, was typical in noting 'the evil smells . . . and an immense quantity of flies'.[22]

The life of the pearl fisher was full of privations. Divers suffered from exhaustion, exposure and perforated eardrums. They fought chest and ear infections, developed scurvy from poor diet, and endured anaemia and persistent skin infections. Tiffany's gemmologist

'Auction of Pearl Oysters in Ceylon', in Edwin W. Streeter, *Pearls and Pearling Life* (1886).

George Frederick Kunz remarked on how many became blind.[23] While sharks weren't commonly a problem (although some attacked and maimed their victims), stingrays, jellyfish, scorpionfish and sea urchins often injured divers. Hallucinations triggered by nitrogen narcosis were common at depth. The men plagued by visions, it was feared, had been touched by a *jinn* (a shape-shifting spirit). The hazards were so great that magicians wove incantations to protect the men from natural and supernatural dangers.[24]

For all the privations, the rewards were scant. The great Moroccan traveller Ibn Battutah, who visited the region around

1330, remarked on the system of credit that left divers perpetually in debt.[25] At the start of the season, the boat captains and financiers advanced the divers cash and goods to sustain their families while they were at sea. When the season ended, the captains arranged the sale of the catch. Following these private negotiations, the illiterate workforce was entirely dependent on the captain distributing fair wages – which never seemed to equal the amount lent to the divers. It was a system of indentured servanthood that bound the workers to their captains and denied them mobility. If they died, their families inherited their debts. At its worst, the perpetual cycles of debt became a form of slavery.

If Gilgamesh's plant was indeed an oyster, an extended metaphor proves it was a bitter prize. While he was resting on his return journey,

> A snake caught the scent of the plant,
> [Stealthily] it came up and carried the plant away,
> On its way back, it shed its skin.[26]

Weeping, Gilgamesh realizes the snake has reaped the fruit of his labour:

> For whom, Ur-Shanabi, have my hands been toiling?
> For whom has my heart's blood been poured out?
> For myself, I have obtained no benefit,
> I have done a good deed for a reptile![27]

These anguished circumstances persisted well into the twentieth century. Villiers observed that the whole rickety economic structure was still built on debt, and divers still poured out their blood and obtained no benefit. He even met young Bedouin forced into diving after inheriting their parents' debts.[28]

The economic exploitation of divers was of such pressing concern that a special 'diving court' was created in Bahrain in 1928. The petitions, signed with fingerprints and annotated with the court's decisions, make wrenching reading. Many involved formerly enslaved divers requesting a *barwa* documenting that they had permission to work as free-divers:

> 'He is, he says, a stranger here, and having lost his own
> *nakhuda* is left stranded without work.'
> 'He says nobody came to engage him as a diver.'
> '*Barwa* cannot be given to this man as he is indebted to
> his *nakhuda*.'
> 'The petitioner is a manumitted slave. He has been given
> a free *barwa* for diving. Now he states that he wants to go
> to Kuwait or Basra to earn his livelihood but he has got no
> money. He requests you to help him.'
> 'No *barwa* can be given to the petitioner. He should go to
> his native country and go diving from there.'[29]

The physical, emotional and financial toll of pearl fishing was paid not just by the crews but by their families and communities. Further south in the Indian Ocean, the Portuguese administrator João Ribeiro painted a harrowing image of the privations suffered by Sri Lanka's poorest families. The pearl fishing season concluded with a fair, where merchants would sort pearls using graduated brass sieves.

> The minute pearls which fall from the sieves are left on the
> sand, and in the rainy season the poor in the neighbourhood
> come to the beach with trays and expose the sand to the air;
> when it is dried by the wind they collect what they call the
> *botica* [the right to winnow sand].[30]

This suffering made its way into the rich pearl-diving musical tradition, but so too did themes of love, courage, hope and religious ecstasy. The music was as diverse as the Persian Gulf workforce – a fusion of Bedouin, Persian, Indian and African traditions, performed aboard ship and when the vessels rested in port. A *nahham* (singer) and musicians performed traditional songs and prayers to motivate, entertain, soothe and inspire:

My soul is like a storming thunder cry
Upon hearing the sound of doves cry
My soul flies from the torments of love
Upon the arrival of *nada* [dew] I cried 'oh my love'
The boat of my passion is lost in a storm
Between tormented waves and the wind's cry.[31]

The torments of pearl fishing were titanic, but so were the profits. For centuries the Persian Gulf was the main supplier of marine pearls to world markets, dwarfing the output of the Gulf of Mannar. Apart from a brief period of contraction in the first half of the sixteenth century, when pearls from the New World flowed like grain into

Rosalba Carriera, *Young Woman with a Parrot*, c. 1730,
pastel on blue laid paper, mounted on laminated paper board.

Bushire (Bushehr), Iran, coastline with *badgirs*, or wind towers,
from George Goudie Chisholm, *The World as It Is* (1883–4).

Europe, the Persian Gulf was recognized as the pre-eminent source
of the highest-quality 'orient' pearls in the world. As the region became ever richer, towns and wealthy port cities burgeoned along
the coast. Anyone travelling by boat in the nineteenth and early
twentieth centuries could have picked out prosperous settlements
by the clusters of *badgirs* (wind towers) stippling the shore.[32] Affluent
Dubai bristled with them. An Iranian architectural tradition, these
slender towers with arched openings, soaring up to 15 metres (50 ft)
above ground, caught cool breezes and drew them downwards into
the living quarters. Some of these structures channelled the air over
subterranean reservoirs, providing natural air conditioning during
scorching summers. Ingeniously engineered from coral for thermal
insulation, they created a pressure gradient to expel stale hot air.[33]
Silhouetted against the violet sky, they signalled the wealth and status
of their owners, many of whom earned their riches from pearls.

Print from *A Memorial of the Marriage of* HRH *Albert Edward Prince of Wales and* HRH *Alexandra Princess of Wales*, 1867, lithograph.

While divers packed together like sardines spread thin mats over oyster-slicked decking, merchants slept in rooms luxuriously cooled by their *badgirs*, each man at opposite ends of the great chain of debt. At the start of each season, the divers were advanced money and food for their families by the captains, who in turn had borrowed from local Arab and Persian merchants. At the season's end, the captains would haggle with these Muslim merchants to secure the highest prices for their catch, settle their own debts and pay their crew. The local merchants were themselves often indebted to the greater pearl merchants based in Dubai, Bahrain and Bombay. At the apex of the food chain, these Hindu or Jain merchants were often closely tied to Bombay, which had emerged as the centre of the global pearl trade by the 1830s. In the first decade of the twentieth century, forty Jain families dominated the Bombay trade.[34] From Bombay, pearls trickled outwards to markets in India, China, London, Paris and the Russian Empire, becoming ever more costly with each continent they crossed.

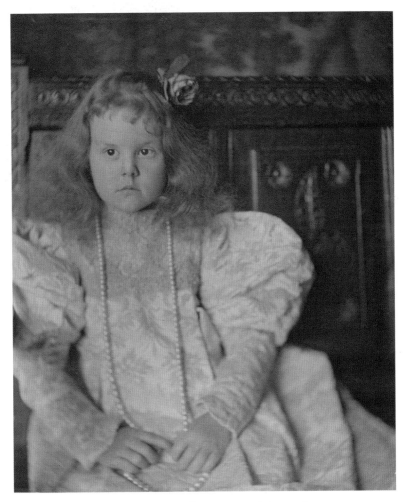

Sarah Choate Sears, *Helen Sears*, 1895, platinum print.

During the last quarter of the nineteenth century, pearls became the jewel *du jour*. They were sported by aristocracy such as Alexandra, Princess of Wales, and American millionaires like Alva Vanderbilt, whose necklace made of more than five hundred pearls was reputed to have belonged to Empress Eugénie and whose tasselled pearl rope hailed from Catherine de' Medici.[35] Aspirational purchases triggered a boom in both exports and prices. Over the 25 years leading up to 1904, gulf pearls more than quadrupled in value. Prices reached a sizzling peak in 1912 when Bahrain exported more than £2,030,000 (today nearly £238 million) worth of pearls.[36]

It was a boom infused with all the hazards of a mono-product economy. João Ribeiro pointed out the problem in relation to Sri Lanka as far back as the mid-seventeenth century. He criticized the over-reliance on pearl revenue, as this diverted people from exploiting the island's other unique resources, such as its cinnamon (reputed to be the world's finest), pepper, elephants and precious stones.[37] In the Persian Gulf 250 years later, if the pearl banks became less productive or were overfished, or if fashions changed and demand

René Lalique, laurel leaves brooch, *c.* 1903, pink pearl, mother-of-pearl, enamel, gold.

dropped, the economic collapse would ripple over the entire area. At the start of the twentieth century, half the region's male population was employed in pearl fishing. In Qatar, the figure was 95 per cent.[38] As the ruler of Qatar, Sheikh Muhammad bin Thani, admitted in 1863 to the British explorer William Palgrave, 'we are all, from the highest to the lowest, slaves of one master, [the] Pearl.'[39] The British, who had tied the emirates in treaties to protect British trade interests in the Persian Gulf, were understandably keen to monitor the fisheries and maintain stability. In his compendious *Gazetteer of the Persian Gulf, Oman and Central Arabia*, the British official John Gordon Lorimer cautioned:

> Pearl fishing is the premier industry of the Persian Gulf; it is, besides being the occupation most peculiar to that region, the principal or only source of wealth among the residents of the Arabian side. Were the supply of pearls to fail, the trade of Kuwait would be severely crippled, while that of Bahrain might – it is estimated – be reduced to about one-fifth of its present dimensions and the ports of Trucial Oman [now UAE], which have no other resources, would practically cease to exist.[40]

It would not, however, be the oyster banks that failed. What crippled the industry was something not yet on the radar of early twentieth-century observers – the slow seep of cultured pearls from Japan. It would be the 1920s before the first warning bells rang. In 1921 British newspapers reported that cultured pearls had entered the market undetected by experts. In the summer of 1924 there were reports of Dubai merchants trying to palm off cheaper cultured pearls as natural.[41] The Bahrain traders, though, felt confident that Japanese farmers couldn't craft large pearls. By 1928 they were more alarmed. The trickle of cultured pearls was having 'a bad effect' on the trade.[42] The Bahrain Government forbade their import, but without specialized equipment, cultured pearls were impossible to detect.

The next year, the United States stock market crashed, triggering a recession that dampened the luxury trade for years. Under these

Pearl diver opening an oyster in front of American servicemen, Bahrain, 1940s, photograph.

twin stressors, the gulf pearl fisheries collapsed because demand rather than natural resources dried up. By 1932 natural pearls were selling for half the price they had in 1928.[43] These were the conditions that sparked the 1932 riot in Bahrain. The desperate divers were deeply in debt to captains who reneged on their loans from merchants who could no longer pay the financiers. Even sheikhs faced ruin. Prices for natural pearls continued to plunge as cultured pearls became ever cheaper, selling for a thirtieth of the price of natural pearls, and nobody could tell the difference. Villiers observed that in consequence, 'Pearls have been cheapened,' and painted a gloomy picture of Kuwaiti pearling vessels lying 'almost as thick as the discarded shells of empty oysters'.[44]

Miraculously, what saved the Gulf economy was the discovery of fossil fuel. Bahrain's first oil well was drilled in 1931, and within seven months, it was producing 9,000 barrels a day.[45] The vast infrastructure – towns, ports, roads – built over centuries to support the pearl fisheries, doubled up to support the booming oil industry, and Bahrain and Dubai, which had grown rich on pearls, continued as

Bahrain pearl merchant Khalil Ebrahim Almoayyed with American servicemen, 1940s.

cosmopolitan emporia. A few *badgirs* remain, dwarfed, of course, by skyscrapers. The Bahrain World Trade Center, completed in 2008, soars 240 metres (788 ft) above the Manama shoreline. Inspired by *badgir* wind-catching technology, its sail-shaped towers funnel the sea breeze into wind turbines installed on three sky bridges. Shimmering above the water, it navigates the small state's twinned currents of past and present, pearls and oil.

CARIBBEAN BLUES

In the Persian Gulf today, banks of undisturbed oysters produce pearls that no one harvests.[46] Off the coast of present-day Venezuela – a site that once pumped pearls in their billions into the global market – the turkey-wing mussel, *Arca zebra*, flourishes where the pearl oyster once thrived. Today cities like Dubai, bristling with skyscrapers, resemble video-game simulacra of the future. On the Venezuelan island of Cubagua in the Caribbean Sea, the ruins of Nueva Cádiz are dismal fragments of the first European city in the Americas. Waist-high stone walls crumble into a sandy plain lapped by the ocean and stippled with *Stylosanthes viscosa* (the aptly named poorman's friend).

It seems fitting that such desolation started by accident. On 3 August 1492 the Genoese navigator Christopher Columbus set sail from Spain, headed for the 'Indies' – the spice-, gold- and pearl-rich lands of China, Japan and India. He was sponsored by the Catholic monarchs Ferdinand of Aragon and Isabella of Castile, whose contract with Columbus promised the explorer a tenth of 'all and whatever merchandise, whether it be pearls, precious stones, gold, silver, spices, and other things whatsoever'.[47] Columbus's confidence in his belief that he would discover pearls was bolstered by his well-annotated Latin copy of *The Travels of Marco Polo*, with its many references to the pearl-rich Orient.[48] On landing in the Caribbean, he saw tantalizing mollusc shells and waters that *might* be conducive to the growth of pearls, but no actual pearls. On his second voyage, he collected boatloads of the wrong type of oyster shell between Cuba and Jamaica.[49] It was not until his third voyage in 1498 that he had

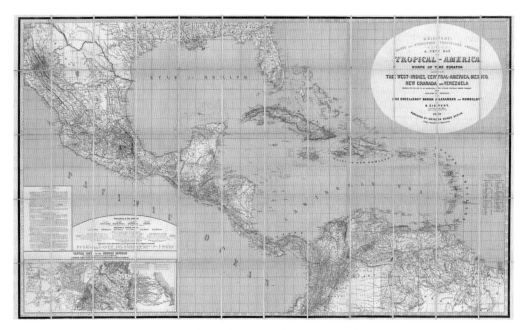

Venezuela's Pearl Coast, detail from H. Kiepert's
Karte des nordlichen Tropischen America (1858).

conclusive proof that the lands he had discovered were blessed with pearls. Passing through the Gulf of Paria, an inlet of the Caribbean Sea between Trinidad and Venezuela, he encountered some pearl-clad islanders, probably local Guayquerí, who told the Spaniards that their gems had come from the north and west.[50] Sailing west, he passed the small islands of Cubagua, Coche and Margarita, soon to become the centre of Spain's pearling empire.

Columbus's tentative discoveries paved the way for future explorers who descended on what, by 1500, was known as the 'Costa de Perlas', or Pearl Coast, along northeastern Venezuela and its mollusc-dense islands.[51] Initially, these entrepreneurs bartered with the indigenous inhabitants, securing pearls in exchange for beads, metals, wine, linen and firearms.[52] As in the Persian Gulf, pearls had been fished in the Caribbean region for millennia. Oysters were a source of nutrition as well as gems, but pearls were much more than ornament. For the region's indigenous peoples, pearls, as containers of light, were sacred. They embodied the life force of the ocean. They were luminous manifestations of spirit.[53]

But for the European opportunists scouring the Costa de Perlas, pearls were commodities. The Spanish Crown attempted to impose a royal monopoly on the trade, a position they abandoned in favour of taxing pearl profits. In 1504 King Ferdinand ordered the Governor of Hispaniola (today the Republic of Haiti and the Dominican Republic) to construct a fortress to protect the pearl trade, and in 1512 authorized private pearling expeditions, to be taxed through the traditional system of the *quinto*, or royal fifth.[54] The *quinto* was more suited to gold or silver, which could be assessed by weight. To tax pearls according to their weight obscured the aesthetic qualities of shape, size and lustre that could exponentially increase the value of any individual gem.[55] It was little wonder that savvy merchants skimmed the best of these small, easily concealed stones to evade the tax.

Some regions along the Pearl Coast, such as the island of Cubagua, were especially fertile. Although prolific oyster beds hemmed this low-lying, arid island, lack of drinking water had long discouraged permanent settlement. This changed during the first decade of the sixteenth century when unlicensed pearlers set up rudimentary shelters for stretches of three to four months during the trading season.[56] Very quickly, supplies of harvested pearls ran dry, and the settlers enslaved indigenous divers from throughout the Caribbean to strip the fecund banks. The Lucayan peoples from the Bahamas, skilled in diving for edible conch, paid a bitter price for their aquatic prowess. Between 1508 and 1518 a population of around 60,000 disappeared from their homeland.[57]

Starved, beaten and shackled at night, the Lucayans sometimes died within days, choking on their own blood from haemorrhaging lungs. The Dominican priest Bartolomé de Las Casas, historian of the Spanish conquest of the Americas, wrote excoriating eye-witness accounts of Spanish brutality towards indigenous peoples. Las Casas, who had sailed from Spain as a teenager, became the conscience of Spanish colonialism in the New World, cataloguing the carnage of innocent people by those 'anaesthetised to human suffering by their own greed and ambition'.[58] The region had become an abattoir not just of indigenous peoples but of the souls of the Spanish, so blackened they could no longer recognize fellow human beings. Spain's New World silver and gold mines devoured indigenous miners, but for Las Casas, the torture of saltwater captives was worse: 'One of the cruellest and most damnable things in the whole of Creation'.[59] Treated as beasts of burden, forced to dive from sunrise to sunset:

> Their hair, which is naturally jet black, takes on a singed appearance more typical of sea-wolves, and their backs come out in great salt sores, so that they look more like deformed monsters than men, or like members of another species altogether.[60]

The colonists treated humans like oysters, as if they were as expendable and replaceable; in consequence, slave prices surged as their numbers dwindled. By 1526–7, the height of Cubagua's pearl production, Lucayan divers fetched up to 150 gold pesos (each peso was worth just over 4 grams of 22-carat gold).[61] Having efficiently decimated the indigenous populations, New World pearlers faced an acute labour shortage, which they solved by importing enslaved Africans. Men and women living in African coastal regions were often excellent swimmers, as water and aquatics were entwined in their belief systems and economies.[62] As demand for divers grew in the New World, slave traders realized the value of African aquanauts, and from 1526 began shipping them to the Caribbean.[63] Nueva Cádiz, the cacophonous heart of the New World pearl trade, was also a centre of the slave trade, the embarkation point for captives, and the nexus through which people were trafficked to the Venezuelan mainland. Initially a ramshackle and lawless village beset by drinking and gambling, in 1528 it was granted the royal title of *ciudad*, becoming the first European city in South America, with a population of nearly 1,000.[64] The stones that now lie in heaps among the shale and scrub were imported from the mainland to construct merchants' houses, a town hall, jail, customs house, granary and a Franciscan monastery with stone gargoyles.[65]

While merchants lived in comfort, enslaved divers' lives were harsh and short. An oceanic trench chilled the water along the Pearl Coast, increasing the danger of exposure. Shark attacks were common. During deep descents, divers risked cardiac arrhythmia and arrest. If they surfaced with empty baskets, they were flogged, which contributed to the beds being stripped of juvenile oysters too young to reproduce or bear pearls. Working in the Caribbean also carried the ever-present risk of encountering pirates who kidnapped and injured slaves.[66]

Nothing about Cubagua was sustainable. As the population swelled on an island of 22 square, unproductive kilometres (9 sq. mi.), the pressure on food sources became crippling. Water, animals, vegetables, flour and fuel all had to be imported. Sheep, goats and cows couldn't survive without pasture, but pigs and chickens scratched by on refuse. Dogs, though, thrived and today still scrounge in skinny packs.[67] Although the oyster beds seemed infinite, within a few years they showed signs of collapse. Over 1,600 kilograms (3,500 lb) of pearls were harvested during Cubagua's most productive year, 1527.[68]

The next year, this figure was slashed in half, and the Spanish king authorized the use of dredges. By 1531, despite restrictions on diving hours and sizes of boats and crew, production continued to tumble. Merchants, used to living well beyond their means, moved on to more lucrative sites along the Pearl Coast such as Margarita Island or further west along the mainland to Cabo de la Vela. By 1540 only a handful of people remained on Cubagua. Those who dug in after a hurricane levelled the buildings in 1541 were soon chased off by French pirates. Cubagua was returned to its stray dogs only 45 years after Columbus's encounter with Caribbean pearls.[69]

Over the short period of their fecundity, the wealth extracted from the Venezuelan pearl fisheries was staggering. Tax records suggest that in under three decades, more than 11,326 kilograms (25,000 lb) of pearls were harvested from Cubagua alone.[70] As tax evasion was notoriously easy, the true figure is doubtless much higher. To reach this production, over a billion oysters must have been pulled from the waters around Cubagua by the hands of enslaved divers; the figure for the Pearl Coast is multiple billions.[71] Despite these numbers, Spain's gain from its pearl fisheries trickled through its economy like seed pearls through a sieve. Pearls were bartered for manufactured goods, and profits dissolved through the settling of debts and foreign wars. By the second part of the sixteenth century, Spain hobbled through multiple bankruptcies. The romantically named Pearl Coast was re-named Nueva Andalucía, and the few surviving indigenous peoples stiffened their resistance.

Some of the pearls at the heart of this parable of greed and loss remained along the Pearl Coast, used for barter or even to buy the divers who lost their lives in retrieving the gems. The rest flowed like grain to Europe and beyond, meeting a craving so intense the era became known as the Great Age of Pearls.[72] During the first half of the sixteenth century, a deluge of fine, lustrous, pale pink pearls from the Americas poured into Spanish ports. In Seville, Europe's central pearl market, they lay trowelled in heaps, 'just as if they were some kind of seed'.[73] Merchants from Nuremberg, Florence, Genoa and Antwerp flocked to Seville, and jewellers, pearl-drillers,

Unknown artist, *Sir Walter Raleigh*, 1588, oil on panel.

entrepreneurs and investors prospered. From Seville, New World pearls streamed across Europe and into Asia. They found their way to royal treasuries, the necks of princesses, the waists of queens, the doublets of queens' favourites, the velvet hats of burghers and the ears of the pirates who ravened the treasure-laden ships departing Caribbean ports. Recorded in painted portraits and inventories, they were given names and assumed the status of celebrities, such as 'The Emperor', comprised of an emerald and an almond-sized pearl; 'The Heart', combining an enormous emerald with a large pendant pearl; and 'The Flowers', made up of a spinel and three baroque pearls.[74] Through the fertile imagination of jewellers, baroque pearls transformed into dolphins, mermaids and lions. And through the

Studio of Hans Holbein the Younger, *Jane Seymour*, 1540, oil on panel.

inventions of apothecaries, pearls were pulverized in *limonada* to cure palpitations.[75]

From the start, a thirst for pearls, precious metals and converted souls spurred the Spanish colonization of the Americas. Appositely, pearls were co-opted by Catholicism, encrusting cathedral altars from the New World to Spain, studding Virgins' crowns and dripping from enamelled crucifixes. One of the many pearls collected by

Joanna of Castile, daughter of Ferdinand and Isabella, was cradled in a small gold urn suspended from her coral and ivory rosary.[76] This single entombed pearl might stand as a melancholy memorial to the entire Pearl Coast catastrophe. Its owner was the child of Columbus's patrons and the instigators of the Spanish Inquisition. Joanna's religious scepticism reputedly earned her the punishment of *la cuerda*, a vicious form of torture where the victim has her hands tied behind her back and is strung up with stones strapped to her feet.[77] Intimidated, disempowered, gaslighted and imprisoned for nearly fifty years, Joanna, like the enslaved divers on the other side of the Atlantic, was the victim of a power-hungry intolerance that denied her personhood.[78] The profligacy of her husband, Philip I 'the

Virgin Mary's Crown, Spanish, 1550–1600, gold, pearl, precious stones.

Crucifixion, northern European, 16th century, gold,
partly enamelled, set with baroque pearl.

Handsome', impoverished her, just as the Pearl Coast was ruined and
impoverished by the time of her death in 1555. Joanna's pearl in its
golden urn mourns life abused and nature ravaged.

MUSSEL MEMORY: FISHING FOR
FRESHWATER PEARLS IN SCOTLAND

According to Julius Caesar's gossipy biographer Suetonius, the promise of freshwater pearls enticed the Romans across the Channel to
invade Britain in 55 BCE.[79] Caesar, an enthusiastic connoisseur of gems

and art, enjoyed weighing individual pearls in his hands, perhaps contemplating whether the aquatic spoils would be worth facing down the belligerent north. In the end, the tribes in the land that is now Scotland proved unfriendly, and the Romans found it easier to source fine marine pearls from the Persian Gulf, the Red Sea or the Indian Ocean. But the idea that it was pearls that changed the course of history – that gave us the cities of London and Bath, decent roads, asparagus and cucumbers, a calendar and currency – is irresistible. We might owe Jane Austen's *Persuasion* to the inconspicuous European pearl mussel *Margaritifera margaritifera*.

Pearl-bearing mussels thrived in northern hemisphere rivers and lakes, from Japan and northern Asia, across Europe and North America. While many species survived millions of years, human industry culled an animal that relied on pristine waterways and was acutely sensitive to heavy metal pollution. Before the Industrial Revolution, some of the rivers that flow through our northern cities hosted prolific colonies of mussels that produced gems of such quality they were compared to Indian Ocean pearls and fooled jewellers into thinking they were 'oriental'. They were fashioned into ropes and stitched onto clothing; portraits of pearl-decked Tudor kings or Hapsburg princesses undoubtedly portray freshwater as well as marine gems. At times, freshwater pearls glutted the markets, but today fishing for these stones is banned in many countries. Small cohorts of pearl mussels survive in remote rivers, but they are long extinct in our industrial and agricultural heartlands.

The tale of Scotland's ruined pearl industry is a narrative echoed around the world but is also a model for hope; this small country – now a global stronghold of *M. margaritifera* – wrought the destruction but sought the restoration of a species that is a barometer of an entire ecosystem's health.[80] Here, in microcosm, is a story of loss and redemption whose lessons ripple outwards.

For countless hundreds of thousands of years, *M. margaritifera* secretly formed pearls in Scottish waters. This long-living invertebrate might survive for two centuries, then release its treasure to be jostled and ground among the river gravel, freeing its calcium to flow to the

Jan Anthonisz van Ravesteyn, *Ernestine Yolande, Princess of Ligne*, 1618, oil on panel.

ocean and sustain other lives. Pearls lie hidden, too, in the streams and small tributaries of Scotland's human history. They are uncovered in museums and archives where they wait, almost as covert as the mussels in which they grew. In the National Museum of Scotland in Edinburgh, on a locket that belonged to the ill-fated Mary, Queen of Scots, pearls twine around the gold-bordered portraits of Mary and her son, James, poignant as milk teeth. And on a ring that belonged to another crushed hero, Bonnie Prince Charlie, tiny seed pearls overlay a lock of his hair.[81] Once, they picked out the initials CR,

Carolus Rex, King Charles, but now these letters are as ruined as his attempt to capture the British throne. In Edinburgh Castle, freshwater pearls glow on the Scottish crown jewels, though the polished gold and other glittering gems tend to steal the show.

'Faire perles' trickle through royal inventories of 'jowellis' (jewels).[82] Handwritten notes on a late sixteenth-century map of Loch Tay in Perthshire mark the presence of watery bedfellows: salmon, trout, eels and pearls.[83] Account books track pearls' export to Europe in the hands of Dutch merchants.[84] Travelogues admire them fastened around the necks of Stockholm's finest.[85] After an enormous pearl was found in Aberdeenshire in 1620, laws stipulated that all pearls fished from Scottish rivers belonged to the Crown, and the biggest and finest

Locket, Scottish, late 16th century, preserved by the Clerks
of Penicuik as a relic of Mary, Queen of Scots, gold and pearls.

Robert S. Duncanson, *Scottish Landscape*, 1871, oil on canvas.

were to be reserved for the king's use. Another Act of Parliament restricted pearl-wearing to the upper classes. The pearl poachers' punishment was 'wairding and laying of thame in the stokkis' – being arrested and tossed in the stocks.[86]

Running alongside these official histories are those of the Summer Walkers, the Gaelic-speaking travelling people of the Scottish Highlands, who were the unofficial custodians of freshwater pearls. They spent the summer months on the road in rural Scotland, living under canvas, selling small household necessities and horses that they had broken in winter to isolated crofters, and fishing for pearls. They fished Scotland's great wildlife rivers – the Spey, the Tay, the Oykel, the Conon, the Naver – and talked of pearls the same way whisky makers talk of single malts. It's the water that crafts the pearl. Its clarity, the minerals and oxygen, and the sunlight filtering down to a clean bed of cobbles all create the river ecology of which pearls and pearl mussels are one small but essential part. 'Show me a pearl and I'll name the river,' boasted one of the last pearl-fishers, Eddie Davies, to the ethnographer Timothy Neat.[87] The river created the pearl, and the pearl mussel made the river.

The Summer Walkers knew exactly which spot in a river was most hospitable to pearl mussels – a stable riverbed of pebbles, larger cobbles and boulders, with fine gravel into which they could burrow. Because mussels, like most of us, prefer company and a little shade, the Summer Walkers scanned the water for patterns of dappled sunlight below overhanging boughs. If this light fell on a 'flat', a current visibly running through a pool, very likely black mussels would be 'sitting there like rows of crows'.[88] Sometimes the water was so clear they could see the mussels on the riverbed, but most often they peered through a glass-bottomed bucket. Pearl fishing relied on sharp eyes, as what they were seeking wasn't the smooth, healthy shells, but ones they called crooks, twisted by the pearl inside. Traditionally, their only other tool was a rod of young hazelwood just under 3 metres (9 ft) long, with a cleft at the end to pluck the mussel from its bed. Opening the mussel with another shell, they might, with the luck of the wise, spy a pale cyst and lift out a pearl, mucous-covered like a newborn.

But the Summer Walkers weren't, by conventional measures, all that lucky. Although they enjoyed 'holiday views', the work was backbreaking and the living precarious. Like most travelling people in Europe, they were sometimes maligned, accused by gamekeepers of poaching salmon or by fishermen of ruining their sport. There were rumours that they stole children. Like most marginalized groups, they kept to themselves and had their own language, a 'cover tongue' that was a variant of Gaelic saved for private conversation.

Beds of *Margaritifera margaritifera* pearl mussels.

Scottish freshwater pearls.

In this language, a pearl was *eanach tom sgaoi*: something big from the water.[89] Most natural pearls are small – occasionally, they would find a pearl large enough to sell in a Highland town for £20. But the optimism of their language kept them going.

The existence of a pearl in *M. margaritifera* is as much of a miracle as the existence of the mussel, as the animal relies on an exquisitely tuned relationship with its environment. In early summer, female mussels inhale the sperm shed by males and for a few weeks nurture the fertilized eggs on their gills. Around August, they release their larvae into the river, ejecting fine white spumes like sighs. The glochidia, the tiny larvae less than 0.1 millimetres in size, must now find a host. Of the 1 to 4 million glochidia a single female mussel can eject over a day or two, almost all are swept away. A few are inhaled by very specific juvenile fish species – the Atlantic salmon and the brown trout.[90] The young fish respond by growing a protective cyst around the intrusion on their gills, a simile for how the adult mussel might later grow a pearl. More passengers than parasites, the glochidia hitch a ride until the following spring, then drop off as juvenile mussels. Picky about their fish hosts, they are just as fickle about their new nurseries, requiring fast-flowing, clean, well-oxygenated water and fine gravel to settle in for the long haul. Like us, they mature very slowly, reaching adulthood at around fifteen to twenty years of age.

An abundance of juvenile mussels of different ages is a good measure of a population's health. Experienced pearl fishers knew to leave the young mussels alone, as they were not yet capable of producing pearls. Nacre is laid down slowly in these cold waters, and a decent pearl might take more than a decade to form. They also understood that these slow-growing mussels were part of a delicate ecosystem and so would patiently wait – sometimes for an entire generation – to return to a colony.

But not everyone who fished for pearls respected the river. Like those of almost all other countries, Scotland's pearl beds periodically collapsed when they were overfished. In the seventeenth century, when European aristocrats slung ropes of pearls around their necks, wound them through their hair and clustered them on silk clothes, vast quantities of Scottish pearls were exported to London. By 1769 the mussel beds were exhausted, and it would be another hundred years before Scottish pearls could tempt buyers again.[91] By then, Lowland factory workers were supplementing their incomes with pearl fishing. This time around, the celebrity status lent by Queen Victoria's fondness for Scottish pearls, combined with water pollution from encroaching industry, effected another collapse.

More recently, William Abernethy's discovery of an enormous, spherical white pearl in the River Tay in 1967 incited another feeding frenzy. Abernethy was rumoured to have been paid £11,000 for the pearl, fondly called 'Little Willie', after its finder.[92] Day-trippers rushed to Scottish rivers to plunder the mussel beds, indiscriminately slaughtering colonies of juvenile mussels. Weekenders with penknives would kill 5,000 mussels for just ten pearls. (By comparison, an experienced pearl fisher could take ten pearls from twenty mussels and then return the animals to the water unharmed.[93]) The banks of remote rivers, newly accessible by car, were littered with shell mounds.

Under current legislation, it is illegal to disturb mussels or damage their habitat, fish for pearls or even possess pearls gathered after

Bartholomeus van der Helst, *Geertruida den Dubbelde,*
Wife of Aert van Nes, 1668, oil on canvas.

1998.[94] But that hasn't eradicated the problem. At least ten times a
year, Scotland's Wildlife Crime Unit is called to investigate piles of
M. margaritifera shells dumped in woodland or car parks or tipped
into rivers.[95] There are very few predators interested in pearl mussels.
The occasional otter or odd hooded crow might be seen plucking live
mussels from their beds and smashing them. An opportunist mink
might spirit a few shells to its burrow. An oystercatcher can stab a

mussel's adductor muscles with its bill and prise out the sweet flesh.[96] But in such quantities, their only real enemies are us.

Pearl mussels are embedded in our written record and illuminate our cultural history, but they do better without us. Our rivers might now be cleaner, but climate change is a present, not future, threat. The weather on the west coast of Scotland has indisputably become more *dreich*. It is measurably wetter and windier, and cloud cover filches more hours of precious sunshine.[97] Some of this rain falls during cataclysmic storms. Floods scour riverbeds and wash out established mussel populations. Juveniles die when silt clogs their gills. Because of their symbiotic dependence on fish, mussels suffer when floods destroy the gravel spawning beds (redds) of salmon and trout. Increased air and water temperature might even uncouple the delicate timing between the pearl mussel and host fish reproduction. The flood defences we build to protect our habitations, like most other river engineering projects, destroy even more suitable habitat.[98]

The planting over the last few decades of non-native conifers such as Sitka spruce, a fast-growing, high-yield crop, might have been good for the economy. But when planted right to the edge of waterways, these trees starved pearl mussels of sunlight and acidified the water. The linear ditches criss-crossing the forests dumped peaty sediment into the streams. But recently, bodies such as Forest Enterprise Scotland, which managed the National Forest Estate covering 9 per cent of Scotland's total land area, have led efforts to replant forests along these rivers and tributaries.[99] Conifers have been replaced with riparian woodland made up of 60 per cent native broadleaves and 40 per cent open habitat.[100] These trees provide the dappled shade beloved of pearl mussels, regulate water temperature and stabilize riverbanks. This benefits downstream aquatic ecology and our own homes and settlements, no longer so prone to severe flooding.

In many ways, pearls – the only gem made by an animal – are the most 'human' of stones, sharing our spaces and timescale. In recent history, each spruce-lined burn and trammelled stream consigned pearls to history or the plaintive metaphors of our poets: Robert Burns weeping 'briny pearls' for lost love.[101] But our rivers can run cleaner within the span of one childhood – the same length of time it takes for a pearl to form. And contained in each one is the weight of a mussel's memory: each *oorlich* winter, each *gloor* day of splintered light, each raw night, each soft summer of no night. The tug of the Moon, the torque of our actions, are locked in nacre layers. The year we built a factory; the year we tore it down. The year we dammed the river; the year we freed it. The year we planted, and the year we pruned. The decades we left the beds undisturbed; the hour we stole the shell from the water. A pearl is a 'minding', a remembrance of an animal secreting the mineral of intimacy.

❀

THREE

Culturing Pearls,
Capturing Markets,
Cultivating Brands

A seasoned pearl fisher could spy a 'crook' and intuit the pearl inside. The logical next step was to invent a more reliable way to stack the odds, priming every mollusc to produce a pearl and averting the waste of animal and human life. Pearl culturing is a refined alliance between human ingenuity and a natural phenomenon. What began as an effort to 'stimulate' nature has evolved into a showcase for technology and innovation. Today's bounty includes chocolate pearls, pistachio pearls, soufflé pearls, pearls carved around gemstones and pearls embedded with microchips that communicate with smartphones. (The Bible, a family photo album or voicemail can be stored in a pearl, literally embodying the scriptural allusion to a pearl of great worth.[1]) Cultured pearls are now shaped by fashion, biotechnology and gastronomical allusion as much as by molluscs.

Running in parallel with the history of fishing for natural pearls is an alternate history of artifice – perfecting techniques to coax a pearl from a living animal. For a long time, that history was a very slender thread, but during the twentieth century, it became the dominant story of how pearls reach the market. With rare exceptions, all pearls available today are cultured pearls, and their appearance depends on both artistry and biology. Bleaching and heating fashioned the 'chocolate pearls' popular in the early 2000s. Decades of selective breeding created Indonesia's golden pearls. Experiments with different nuclei

Blister pearls in the form of Buddhas, Guangdong, China.

Blister pearls in the form of Buddhas, Guangdong, China.

have produced coral-pink pearls, and puffy soufflé pearls are cultured by inserting mud into a harvested pearl sac. Perhaps the most dazzling artist's intervention is Chi Huynh's use of amethyst, turquoise or citrine nuclei beads. He carves the surrounding nacre to reveal glimpses of the inner jewel, like mace around nutmeg.[2]

Humans have been making pearls for centuries. Or trying to. The first attempts might date from the fifth century CE in China, where objects such as small Buddhas or fish were inserted into freshwater mussels to form blister pearls.[3] The process fascinated foreign visitors, who described it in travelogues and letters home. By the mid-nineteenth century, the art flourished in Zhejiang Province in eastern China, a region still at the centre of the global cultured pearl industry. Following a centuries-old tradition, children gathered the mussels in spring and wedged them open with bamboo. The adults performed the delicate surgery, slipping brass, bone, pebbles or mud between the shell and the mantle tissue. The mussels were returned to ponds, nourished with night soil (human waste) and retrieved three years later. Millions of these doctored shells flooded the markets of nearby Suzhou, fetching pennies a pair.[4]

The first cultured spherical pearls, though, were created in Europe. A necessary step was understanding that shells and pearls are the products of biomineralization, not a romance between an oyster and a moonbeam. The eleventh-century Persian scholar al-Bīrūnī observed that a pearl consisted of concentric layers of the same substance as the inner surface of the mollusc's shell. In 1609 the Flemish mineralogist Anselmus de Boodt conjectured that the pearl grew from fluid secreted by the animal, which hardened around a core. The invention of the microscope permitted a closer look at the pearl's onion-skin layers, as famously described by the French scientist René Réaumur in 1717. He suggested that the process was triggered by disease, an idea that had occurred centuries earlier to medieval Arab scientists.[5]

It wasn't just intellectual curiosity that led entrepreneurs to intervene in a natural process, though. From the start, pearl culturing, or periculture, was connected to regional or national economic development. This played out, first in failure, then in international success, in the very different contexts of Enlightenment Sweden and modern Japan. But economics wasn't the only driver. Periculture has scientific, political and ethical dimensions informed by our beliefs about our relationship to the natural world. The industry navigates the often-contested ground between ownership and stewardship, extractive capitalism and sustainability. It takes millions of animal lives each year, and at times it has trampled on the rights of indigenous peoples, yet offers livelihoods in developing countries. It is dependent on technology and intellectual property, but also on art and accident. It promises accessibility, challenges the authority of connoisseurship and tradition, while upholding its product as a luxury. In short, pearl culturing is riddled with the contradictions of many of our partnerships with nature.

LINNAEUS IN LAPLAND

The first person to culture a spherical, rather than blister, pearl was tidy-minded Carl Linnaeus, the Swedish botanist much better known as the Father of Taxonomy than for his efforts with mother-of-pearl. A second Adam, he scrapped the haphazard naming attempts of the first, dividing each organism into genus and species: *Homo sapiens*, *Panthera leo* (lion), *Margaritifera margaritifera* (pearl mussel), *Nautilus pompilius* (chambered nautilus – also known, untidily, as the pearly nautilus, the emperor nautilus, Nautile flammé, Lagang: the problem Linnaeus mopped up with Latin).

He believed pearls, not taxonomy, would make him rich. Though not by fishing them from the likely, though overfished, places – the Persian Gulf, the Red Sea, the Gulf of Mannar – but by making them an act of creation as immodest as his ordering of the world. For a time, it seemed to his colleagues that he would be stolen from botany. One joked, 'I am afraid you will be spoilt for a gardener you will grow so rich with breeding of Oriental perls.'[6]

It wasn't a trip to the Middle East, though, that lodged this idea in his mind, but one to a region now tied closer to Father Christmas than to the Father of Taxonomy: Lapland. In this landscape – where he was fond of noting that nothing much happened – plans for pearl culturing and new botanic ventures on the tundra began crystallizing in his mind like the snow that persisted at elevation year-round.

On 12 May 1732 the 25-year-old son of a provincial vicar set off for Swedish Lapland from the university town of Uppsala, just north of Stockholm. He was jauntily dressed in a wool and linen coat, leather trousers and a green cap with earflaps over his wig. In the countryside around Uppsala, the barley had just sprouted, and there were fresh green leaves on the birch, elm and aspen trees, but Linnaeus was prepared for a far less gentle climate. Armed with a gauze midge hood, microscope, magnifying glass and journal, he anticipated flying pests and a botanical bonanza, but probably not the dream of pearls that was to preoccupy him through his next thirty years.[7]

His journey would take him 3,200 kilometres (2,000 mi.) on foot, horseback and boat around the Gulf of Bothnia, with forays into the remote interior of northern Scandinavia. Swedish Lapland at this time included parts of present-day Finland and Russia and stretched almost as far north as the Arctic Ocean. Linnaeus would be gone until October, funded by a grant from the Uppsala Royal Society of Arts and Sciences. The trip was a chance to develop his ideas about plant classification in an 'exotic' locale and discover medicinal plants used by the indigenous Sami people.

The main purpose of this trip, though, was not to sketch a botanic portrait of Lapland but to identify resources the motherland could harvest. Linnaeus's Lapland adventure was an economic foray infused with Enlightenment dreams of territorial development. Lacking an overseas empire and having lost the Baltic lands it had gained the previous century, Sweden was thrown back on its own reserves. From the early seventeenth century, it had pushed northwards to the Arctic Ocean, across territories used for centuries by the Sami for reindeer herding, fishing and hunting.[8] These incursions opened new trade routes and also development opportunities. As a civil servant, Linnaeus shared his generation's drive for economic self-sufficiency and alarm that Europe's silver was draining into the 'sink' of China in exchange for tea, silk and luxuries.[9] For visionary Linnaeus, the answer lay in producing tea, silk and pearls *in* Sweden.

Part economic spy, part botanist, part ethnographer, Linnaeus meticulously recorded his trip in his travel journal, now owned by the Linnean Society of London. Linnaeus's firm, forward-slanting hand-writing wraps around his sketches of plants, animals and tools used by the indigenous Sami people. He dutifully records the locations of silver ore and iron and where he collected samples of quartz, slate and petrified coral. He notes which rivers host vast runs of salmon, which earths are useful for pigments, which mosses dye yarn. But this journal is much more than a catalogue of plants, animals and minerals that might benefit the motherland. It's a long, hard look down a microscope.

As if allowing time for his eyes to focus, he describes 'nothing much worthy of notice' at the end of a day in Luleå, high up in the Bothnian Gulf: nothing much, except for the sea lashing the road and nibbling islands from the landmass. And stretches of marsh with cranberries and bilberries blooming red and pink. Mosquitos hung in low clouds over damp meadows. At sunset, he took a ferry over a broad river. On the road into town, he found a gibbet with the headless bodies of two Finns executed for highway robbery and the quartered body of a Laplander who had murdered a relative. In bed that night, he was startled awake by a blaze of fire on his wall and realized it was the midnight sun rising.[10]

There was nothing to be got in Luleå. Except a type of bristly grass called Old-man's beard in southern Sweden, Hog's bristles fur-ther north, and in Luleå, Lapland hair. Linnaeus named it *Nardus stricta*, tight grass. In this one species, there was a whole world of naming, vernacular bumping up against Latin, indigenous against imported.[11] It was the same with the calendar. The Sami, coercively Christianized for centuries, punctuated their year with Midsummer Week, Goose Week, St Margaret's, St Olaus' Mass, Reindeer Fawn Week, Holy Cross and St Matthew.

From Luleå, Linnaeus headed far inland towards Norway. He pocketed silver, iron ore and mica. He travelled through dense birch and pine forests, home to black, blue and white foxes whose glossy coats realized fortunes. These pelts of fox, ermine, lynx and brown bear had enticed Swedes to Lapland. By the 1630s they realized that the land harboured much more than fur-bearing animals, and soon silver, iron and copper mines, partly serviced by conscripted Sami labour, bolstered Swedish coffers.[12]

Although Linnaeus projected an image of himself as a rugged solo adventurer, he depended on his Sami guides, who interpreted for him, sold him food, leased him pack animals and entertained him. The success of any future ventures, including pearl fishing, would rely on their co-operation and labour. In his diary, the Sami drift in and out of focus, sometimes invisible, sometimes rendered in microscopic detail. Each morning, thousands of their reindeer returned home to be milked, and his Sami hosts offered Linnaeus boiled reindeer milk that had the texture of cows' milk mixed with eggs. He watched Sami children play with birch twigs fashioned into antlers and babies bound snugly into reindeer leather cradle carriers. He saw lakes bloom white with water ranunculus (*R. aquatilis*), noticed how gracefully reindeer swim and raptured over the pristine emptiness and dazzling, snowy waste.

It was, of course, convenient to find Lapland empty, to find 'nothing'. Emptiness invites plunder. For all his enthusiasm and fascination with the Sami, Linnaeus was an Enlightenment explorer, eager to profit from what he encountered. A hard-nosed negotiator, he traded tobacco for reindeer cheese double its worth. He romanticized the landscape and the Sami, a people decimated by disease, alcoholism and subsistence living, and overlooked their hardships: 'I never met with any people who lead such easy, happy lives.'[13]

The higher he climbed, the more stunted the trees. One lake was the colour of water poured into a bowl previously used to hold milk. Approaching a pass, he noted there was 'nothing' except pristine lakes below ridges of mountains rising like shoulders.[14] And at tranquil Purkijaur, he found a wooded oasis not far from a pearl fishery.

> We arrived at length at Purkijau [*sic*], a small island, the
> northern side of which is planted with forests of spruce fir,

and the others with woods of birch, by way of protection
to the corn. The colonist who resides here informed me
that the corn never suffered from cold.[15]

A month later, after looping westwards over a white landscape
that seemed to vibrate like ripples on water, he returned to Purkijaur
to take a closer look. He had, perhaps, been misinformed about the
mildness of the climate. On 26 July he arrived back in this region of
clear, gravel-bottomed lakes fed by fast, foaming rivers. Fog made
the evening unusually dark, but Linnaeus was determined to press
onwards. He rigged up a small raft and punted out into the current.
It was a dangerous misjudgement. Before long, the force of the water
shredded the flimsy craft, sweeping away a valuable cache of botanical
specimens. Clinging to the remnants of the raft, he struggled to an
island half a mile downriver.

Admitting his hubris, he hired a guide to show him the mussel
beds. In his journal, he sketched the rudimentary raft his guide made
for him, equipped with a pole and a stone on a cord for an anchor.
In the drawing, a small bearded figure lies prone, peering through
the water to the bed of boulders and a few of the mussels Linnaeus
was to name *Mya margaritifera* (now *Margaritifera margaritifera*).

Purkijaur had been a productive pearl fishery before Linnaeus's
visit. He heard rumours that, in the past, the mussels clustered so
thickly on the riverbed no one could reach the bottom. Now it was
'nearly exhausted', and its caretaker had to make a new raft for him.[16]
During the sixteenth and seventeenth centuries, Sweden's freshwater
pearl fisheries were productive and lucrative. From the late seventeenth century, the Swedish Crown had a monopoly on pearl fishing
and appointed overseers. During Linnaeus's first visit, he mentioned
a colonist who lived near the fishery, an oblique reference to that
other army of people on whom Linnaeus relied during his trip for
hospitality and guidance. Government incentives lured Finnish and
Swedish farmers to settle in Lapland through land leases, hunting
privileges and tax breaks. Linnaeus's colonial hosts – parsons, mine
superintendents, schoolmasters and judges – represented 0.2 per cent

Unknown artist, *Icon of the Descent into Hell*, Russian, frame 18th century,
icon 19th century, tempera on wood, metal gilt, painted enamel on copper,
freshwater pearls, mother-of-pearl.

of Lapland's people, but their administration governed 40 per cent of the country's landmass.[17] Lapland's pearl fisheries may have been remote, but they were wild only in the sense of a wilderness tapped for improvement by the non-indigenous.

But for the Sami who had lived in the region since at least the Viking Age, pearls were supernaturally radiant, and rivers and land were spiritually animated.[18] Pearls, water and precious metals were not resources to monetize. Sami involvement with the pearl fisheries started when the Swedish colonizers encouraged them to hunt for minerals and resources while they herded reindeer through remote regions. In the decade before Linnaeus's visit, the administration proposed that pearl fishing could alleviate Sami poverty.[19] The agriculturalist Swedes were busy farming in the summer, leaving the rivers free for the Sami to fish, but there was sometimes conflict, or at least competition, over access to rivers.[20] Seasonal pearl fishing supplemented livelihoods, especially since the Russian Orthodox Church prized pearls so highly and created a reliable market.

At Purkijaur, Linnaeus watched from his raft as his guide pulled up a mussel with a pair of wooden tongs and used a whelk shell to 'thrust with violence between the valves'. It disturbed Linnaeus that thousands of mussels were destroyed to find just one pearl. He noted that their white or reddish tints related to the colour of the nacreous inside shell. The seed of an idea planted itself in his inventive mind. Glittering Purkijaur was a magical place. He saw an entire duck retrieved whole from the belly of a pike.

On his return to Uppsala, Purkijaur and the economic bent of his Lapland trip continued to play on his mind. He had read about Chinese blister pearls and thought that the Chinese method was so straightforward it was astonishing more Europeans weren't attempting it. The Director of the Swedish East India Company, Magnus Lagerström, sent Linnaeus some of these pearls as part of a haul from Asia that included flowers, cotton seeds, Chinese grasses, tea plant seeds and tropical fruit from Java.[21]

In 1748 he revealed to a fellow botanist that he could tell the time from the opening and closing of a flower's petals and that he had invented a method to make all mussels bear spherical pearls. This endeavour, clothed in secrecy, would allow him to enjoy the good things in life instead of labouring as a scientist.[22] Initially, Linnaeus envisioned generous cash grants from the government, a common economic stimulus.[23] He dreamed of an inexhaustible supply of pearls enriching both Sweden and himself. His Lutheran upbringing had been without frills; as a student, his access to resources, even books, had been meagre; as a professor of medicine at Uppsala University, he moonlighted as a tour guide, until the rector of Uppsala University clamped down, causing Linnaeus two months of sleepless nights. He knew he couldn't rely on the scant harvest of academia to support the lifestyle he craved.

His relative penury and perception of how easily pearls could be cultured explain the paranoia that creeps through his correspondence. He knew that many fast-flowing northern European rivers hosted pearl mussel populations, and he was anxious that 'foreigners may not obtain knowledge'.[24] Nature, knowledge and nationalism coalesced. He feared that if his secret were made public, cultured pearls would flood the market and their prices collapse, washing away his dream of a leisured retirement as swiftly as the river at Purkijaur had swept away his plant specimens.

Remembering the sight of opened mussel shells littering the river banks at Purkijaur, he began to dream of creating a large commercial enterprise there. The mussels could be harvested, and three men could operate on a hundred a day. His own experiments taught him that a skilled technician was essential. A gem-quality pearl 'depends on the person who impregnates the muscle, to cause her to bear a ripe or unripe pearl'.[25] The mussels could then be returned to the river in cages to wait for nature to perfect the process. He knew that nacre accumulated slowly in these icy waters, and he reckoned it would take six years to grow a pearl the size of a pea and twelve for one double that size.[26] As for labour, he didn't elaborate on the human capital required to operate this venture. The Sami, presumably, would keep the farm provisioned. He had envisioned other Lapland ventures worked by children and the disabled. Possibly he thought a variety of marginalized people might provide a supplementary workforce for the fishery.

Far inland from the coast, Purkijaur's isolation made it an ideal setting. From Luleå, it was a long journey by boat and on foot. From Norway, the trip over the mountains was even more arduous. Any prospective spy would require the help of Sami guides. Linnaeus was satisfied that he wouldn't be the victim of industrial espionage.

The Purkijaur pearl fishery was just one of Linnaeus' visions for the country's prosperity. With the benefit of hindsight, so many of his dreams seem like snow blindness. He believed Sweden could have a home-grown silk industry, and he imported mulberry bushes to feed the silkworms. The bushes froze. He hoped to dilute the China tea trade by establishing tea plantations. The plants failed to thrive, even in the milder south. He proposed saffron, nutmeg and cinnamon plantations in Lapland, bolstering the population and increasing tax revenue. But he couldn't harden sensitive plants like coffee, sugar and ginger to the cold, and Lapland never experienced its influx of spice farmers. Nature may well have been 'an infinite larder', but a country's climate and geography dictated the arrangement and bounty of its shelves.[27] Even with diligent husbandry, tea, coconuts and cloves lay far out of reach.

Pearls, though, should have been different. *M. margaritifera* thrived in pristine northern European rivers. The economist in Linnaeus was appalled that so many mussels had to be sacrificed for just one gem-quality pearl; the theologian in him rued the squandering of a God-given resource. To 'breed' or culture pearls would be a redemptive act of stewardship. Trained as a doctor, he was inclined to diagnose a pearl as an illness in the animal. The illness in the pearl-fishing economy could be cured by culturing the pearl.

His experiments with pearls were more successful than his efforts to acculture tea. In Uppsala, he only had access to the river mussel *Unio pictorum*, which he gathered from a slow-flowing urban creek. He drilled a small hole through the shell and implanted a tiny limestone bead between the mantle tissue and shell, held in place by a silver wire that prevented the bead from growing as a blister pearl.

Linnaeus's experimental pearls cultured from the freshwater mussel *Unio pictorum.*

He was approached by the Austrian naturalist and ambassador Joseph von Rathgeb, who wanted to buy the rights to his invention, but Linnaeus was holding out for a bigger reward.[28] In 1761 he explained his technique to a subcommittee of the Swedish State Council. When nine of his cultured pearls were sawn in half as proof of his method, the royal jeweller confirmed them indistinguishable from natural pearls. The committee recommended a reward of 12,000 silver dollars. Sadly, the mechanisms of bureaucracy worked slowly, and when the transaction finally took place the following year, Linnaeus received only 6,000 dollars.[29] As a sweetener, he received a title, ennobling him in name if not in purse.

The person who purchased Linnaeus's secret was the Gothenburg merchant Peter Bagge. There is no extant evidence that Bagge did anything with it, and the invention lay hidden until two generations later in the early nineteenth century when Bagge's grandson, Jacob, was sorting through his grandfather's papers. In the spirit of missed opportunities, Jacob, too, failed to exploit Linnaeus's discovery.

A small collection of his pearls survives at the Linnean Society of London, some still attached by wire to the shells, mutely sharing his secret. The tiny collection reveals in microcosm the hazards and disappointments of pearl culturing. Some are chalky brown, some like scaled seeds or teeth or the bodies of insects. But four glow, refracting light like the Lapland ice.

MIKIMOTO IN MIE

Pearl culturing progressed in earnest a century and a half after Linneaus's experiments. Kōkichi Mikimoto, whose name is now synonymous with cultured pearls, began business while still a teenager in the 1870s as a vegetable dealer in Toba, a small Japanese port town in Mie Prefecture. Barred from selling his wares to British sailors on their warship, he put on a juggling act from his rowboat, spinning eggs and potatoes above his head as deftly as he would later juggle the complexities of a pearl empire. At seventeen, showmanship won him access to a captive crew craving fresh produce. As the 'Pearl King', who appreciated the optics of trade as well as jewels, he would capture new markets for affordable pearls around the globe. But as with juggling, the ease with which Mikimoto built his pearl empire only *looked* nonchalant. Dig deeper, and it's clear Mikimoto spent decades walking on eggshells. While the story of Linnaeus's pearls can be told through his Lapland journal and letters, that of Mikimoto, a man of spectacle, is revealed through images.

Measuring pearls for Amaterasu

In later life, Mikimoto was photographed weighing pearls in a balance, like a twentieth-century Vermeer genre painting. Another pile of pearls gleams against a dark mat resting on a lace tablecloth. He has the same intense focus he had as a boy juggling eggs. The photograph's caption reveals that Mikimoto is, with infinite care, measuring pearls to offer to the Grand Shrine of Ise treasury.[30]

This shrine on the Shima Peninsula, dedicated to the sun goddess Amaterasu, is one of the most sacred sites of Japan's indigenous Shinto culture. It lies about 16 kilometres (10 mi.) inland from the port town of Toba ('no larger than a narrow sash'), where Kōkichi Mikimoto was born in 1858 to a working-class family who owned a noodle restaurant.[31] Decades later, he was a national hero and purveyor of pearls to the imperial household. That he was preparing a gift of prized pearls for the shrine at Ise, home to the mythical ancestor of Japan's emperors, suggests that pearls held an important role in the country's traditional material culture. It was a useful marketing tactic to encourage buyers to think of pearls as Japanese, just as porcelain was Chinese, tea was Indian, whisky was Scottish and fashion was French. This idea has been resilient. As Japanese singer and style icon Mari Natsuki recently reflected, a pearl

Kōkichi Mikimoto weighing pearls to be gifted to the treasury
of the Grand Shrine of Ise, dedicated to Amaterasu.

'reminds us of our Japanese identity . . . It makes you feel Japanese and makes you feel proud'.[32]

But widespread pearl-wearing in Japan is recent and was boosted by Mikimoto's innovations. Historically, pearls weren't broadly favoured in Japanese dress or jewellery, which surprised foreign visitors such as the seventeenth-century French gem merchant Jean-Baptiste Tavernier, who bluntly concluded that the Japanese 'do not esteem pearls'.[33] Compared to Europe, pearls weren't prominent in a religious context either. Shinto reveres spirits residing in nature rather than in objects, so there was little call for pearl-studded altars or liturgical art. And while pearls function symbolically in Buddhism, a religion imported into Japan by the sixth century CE, this didn't translate to a fashion for pearl artefacts. By comparison, in the Byzantine Empire (330–1453), the Church used pearls even more lavishly than the imperial court, crusting icons and chalices with the gems and burdening church vestments with seed pearls.

But in Japan, the use of pearls in Buddhist temple treasuries is more restrained. The oldest treasuries are stocked with rarities traded along the Silk Road – gold, crystal, amber, ivory, nautilus shell and rhinoceros horn. But few spherical pearls. The most famous historical treasury, the Shōsōin of Tōdai-ji in Nara, has a well-preserved collection of 4,158 Akoya and abalone pearls dating back over 1,200 years, but most of these adorned the emperor's crown.[34] The others are in ruined fragments of headdresses or attached to textile scraps or set into dagger handles. The devout Emperor Shōmu (701–756), who owned many of the treasures in the Shōsōin, had a pair of red leather slippers modestly studded with small pearls.[35] At nearby Kōfuku-ji, priests offered a handful of pearls to the earth spirits to ask permission to build the new 'temple that generates blessings'.[36] Mother-of-pearl – the iridescent nacre that coats the inside of pearl-bearing molluscs' shells – was far more widely used. Ninth-century Japanese literary sources mention lacquer-work inlaid with mother-of-pearl.[37] It was used to sumptuous effect in the interiors of temples dating from the eleventh and twelfth centuries, and on boxes to store sutras, musical instruments, saddles and boxes.[38] The use of

Box and cover, Japan, late 19th century, *shibayama* technique carved
mother-of-pearl and tinted ivory, silver, lacquered wood.

mother-of-pearl as a decorative material continues to be an art form
that has been perfected in Japan over centuries.

The earliest written records of Japanese pearls come from
Chinese sources in the first to third centuries CE, reflecting their
value in China.[39] Later Japanese records reveal pearls were given as
diplomatic gifts or traded with Chinese merchants. China's continued
demand for Japanese pearls and marine products inspired Mikimoto
to begin his experiments improving the quality and yield of marine
animals. Qing dynasty China had an unquenchable appetite for edible
kelp, abalone and sea cucumber, harvested by Japanese divers and
fishermen. These products were dried, sold in bulk at Japanese ports
and shipped to coastal Chinese cities like Ningbo. While European
nations worried about the drain of silver to China, Japan had found
a way to turn the currency tide with its profitable marine exports.

When he was nineteen, Mikimoto accompanied a Toba merchant
to the port of Yokohama and saw this lucrative trade for himself.

The teenage entrepreneur also observed the trade in seed pearls that Chinese merchants bought in bulk, unconcerned about their quality. He had stumbled upon the use of pearls in Chinese medicine to restore eyesight, dewy skin and virility. It was rumoured that even the opium smoker's constipation and itching skin could be eased with pearl powder.

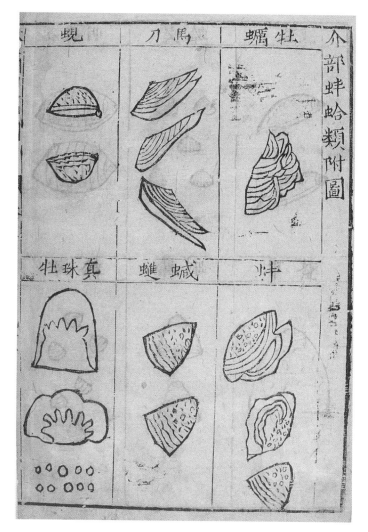

Li Jianyuan, illustrations of shellfish including pearl oyster in Li Shizhen's pharmaceutical encyclopaedia *Bencao Gangmu* (*Compendium of Materia Medica*), 16th century, woodcut.

However, by the 1880s, when Mikimoto was expanding his marine products business, overfishing had decimated stocks of oysters, shellfish and other edible marine animals. Concerned fisheries officials and zoologists began exploring how to conserve populations and enhance breeding. Mikimoto's success at improving the processing of sea cucumber brought him to the attention of the Fisheries Society of Japan.[40] His path to measuring pearls for the sun goddess at Ise began with the leathery sea slug, prized for its slimy, slightly crunchy meat that soaks up seasonings like a sponge and as medicine for impotence, poor kidney function and constipation.

Oysters' Souls

A throng of schoolchildren watch Buddhist monks chant sutras in front of a tower of oyster shells.[41] It looms above them like a miniature Mount Fuji. They are among the thousands of witnesses at a memorial service Mikimoto held in October 1936 to atone for the animals sacrificed for his pearl business.

Kōkichi Mikimoto and mound of oyster shells, memorial service, 1936.

Mikimoto, Sakura sash clip, 1908–12, 15-carat gold and platinum setting, in a traditional Japanese cherry-blossom pattern, set with a cultured half-pearl in the centre.

> To date I have killed 150 million pearl oysters . . . Holding this memorial has lifted a weight off my chest. I can now enter the Land of Ultimate Bliss (*gokuraku*) without hesitation. But my goal is to live to the age of eighty-eight. I will kill 500 million oysters by the time I die.[42]

He made sure his contrition was public. On the left side of a photograph that captured Mikimoto's declaration, a cameraman records the moment. The news was picked up by the Associated Press and broadcast internationally.[43]

At the start of his career, Mikimoto focused less on killing and more on mollusc cultivation, reproduction and repopulation. In 1890 the prominent zoologist Kakichi Mitsukuri noticed his success and invited him to the Marine Biological Station at Misaki.[44] Inspired by his week studying pearl formation with Mitsukuri, he began experiments surgically implanting different nuclei into living oysters.

It is easy to admire the glossy surfaces of empty seashells without considering the living creatures that once inhabited them. As aesthetic objects, shells, like tombs, exist apart from the organism. The pile of empty shells in the 1936 photograph, groomed like one of the popular ornamental Fuji mountains that graced Japanese gardens, is an image cleansed of slaughter.[45] The Buddhist ceremony focused on the spiritual, but the business of pearl culturing, in which animals are surgically manipulated and later harvested, is entirely corporeal.

Pearl oysters and mussels have simple nervous systems, even in comparison to other molluscs such as octopi. Based on what we know

Ago Bay, Mie Prefecture, Japan, aerial photograph, 2015.

of their neural circuitry, they are unlikely to feel the emotional and affective components we recognize as pain.[46] But the experience of a species so radically different from us is unknowable; our recognition of other species' sentience is limited. While the oyster's surgery may not be 'painful', it can be physiologically stressful and cause the animal's death.

Even if suffering is absent, the cultivation and killing of animals for profit was an ethical inflammation the Buddhist ceremony was intended to salve. Using animals as biocapital – commodifying life – depends on a worldview that places humans outside and above nature. In a publicity pamphlet Mikimoto produced in 1907, he explained that his aim was to 'make the pearl-oyster work for man and produce natural and true pearls in a reliable and methodical manner – in short, a kind of "harnassing" the mollusc for the service of man'.[47] The metaphor suggested the surgical process was no more troubling than hitching a pony to a cart.

After studying marine biology with Mitsukuri in 1890, Mikimoto began modifying the Chinese technique for culturing blister pearls. He settled on mother-of-pearl nuclei beads made from freshwater mussel shell but grappled with his oysters' tendency to spit out the beads. He fought against octopus partiality to his molluscs and against the far more devastating scourge of 'red tide', an algal bloom that suffocated them in their millions. With persistence and skill at hiding from his creditors, he achieved a milestone in 1893. He had cultured a hemispherical pearl like a nipple attached to the shell, for which he received a patent in 1896.

Although they weren't the coveted spherical pearls that could be strung into expensive necklaces, hemispherical pearls were a valuable cash crop. Flat-backed, they could be mounted on earrings, rings and brooches. To manage his oyster stocks, Mikimoto began marine farming on an industrial scale, leasing an island and its surrounding waters in Ago Bay. About 20 kilometres (12½ mi.) south of Toba, Ago Bay's nibbled coastline, sheltered islands and inlets harboured large stocks of Akoya (*Pinctada fucata*), the pearl-bearing oysters.[48] Mikimoto bought juvenile oysters from local villagers and raised them to adulthood, assisted by female divers he hired to tend his 'crop'. From 1896 he was operating on 250,000 oysters per year. By 1902 this figure rose to a million.[49] By 1907 he claimed to command 29 nautical miles.[50]

In leasing this area for his first aquatic farm, Mikimoto was plunging into the contentious waters of fishing regulations and access rights to marine animals. The ethics and politics of this were to dog him for decades. Although Japan's coasts were state-controlled, individual villages could secure licences to cultivate the waters along their boundaries and access off-shore waters. These arrangements were fluid but not uncontested. As Mikimoto's farm expanded and he pressed for renewal of long-term leases, he was inevitably drawn into conflict with villages and other marine entrepreneurs eager to profit from these sheltered waters.[51]

For most of his career, Mikimoto steered a difficult course. The looming pile of oyster shells photographed in 1936 was evidence of Mikimoto's animal dilemma – and that of industrialized Japan. An oyster is a natural resource with the potential to bear a pearl. In Buddhist thought, it is also a sentient being with moral and spiritual qualities, capable of a better rebirth. Periculture, like deep-sea whaling, brought modern Imperial Japan prosperity and prestige. But as Mikimoto and whalers discovered, these industries also attracted censure. Like Japanese memorials for whales the previous century, Mikimoto's services for oysters were spiritual interventions designed to restore the balance between nature and human need.[52]

'Pearls and Jewels, K. Mikimoto'

In June 1907 an elegant advertisement appeared in the *Japan Times*, the country's most important English-language newspaper. Cradled on a half shell are the words 'PEARLS and JEWELS, K. MIKIMOTO, No. 3, GINZA-SHICOME, TOKYO'. Pearls as big as soap bubbles rise from the surrounding ripples. Two years previously, Mikimoto had achieved the holy grail of pearl culturing – his first round 'free' pearls. It was a singular accomplishment. In early 1908, a few months after the appearance of this advertisement, he received patents protecting

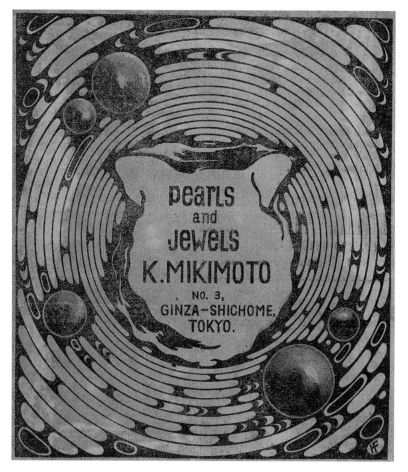

Advertisement for Mikimoto's Ginza store, *Japan Times*, June 1907.

his discovery. However, it would be 1916 before he could start culturing high-quality round pearls on an industrial scale and the 1920s
before they were widely available.

Even though 'Mikimoto' is the name on the advertisement and
above his shop on elegant Ginza, the most fashionable shopping street
in Tokyo, his achievement wasn't entirely singular. The mechanism by
which oysters make pearls involves more than an irritant introduced
as a nucleus. In natural pearl formation, some of the nacre-secreting
epithelial cells from the outer layer of the mantle are carried along with
the foreign body and grow to form a protective pearl sac around it.

Understanding the vital function of the pearl sac was central to designing a reliable method to culture pearls. In the early years of the twentieth century, Mikimoto collaborated with the dentist Otokichi Kuwabara, who modified the tools of his trade to make chilling instruments for oyster surgery, some of which are still used today. Together they devised a cumbersome method that owed much to Kuwabara's suturing skills. A nucleus was sewn snugly into a piece of mantle tissue harvested from another oyster and implanted into a new host oyster.

Two men with a better understanding of the function of the pearl sac independently designed a less time-consuming method.[53] Their discoveries helped Mikimoto scale up production, but the latter's entrepreneurial brawn overshadowed their names. The carpenter Tatsuhei Mise invented a needle that could inject a tiny nucleus and a piece of epithelial tissue into the mantle of an oyster. He received a patent for his needle in 1907. The other major figure was marine biologist Tokichi Nishikawa. Nishikawa visited Mikimoto during an outbreak of red tide and met Mikimoto's daughter, Mine, whom he married in 1903. However, his relationship with his father-in-law was

Nucleus beads and strips of mantle tissue used in pearl culturing,
Mikimoto Pearl Island, Toba, Japan.

tense, and he continued to conduct his own laboratory experiments and fieldwork. He discovered that implanting a piece of epithelial tissue with a nucleus stimulated the creation of the pearl sac – essentially the same discovery made by Mise the same year, and the process still used today for saltwater cultured pearls. After months of acrimony, with Mise and Nishikawa both insisting they had crossed the finishing line first, they signed an agreement in 1908 to become co-owners of the new technology, the Mise–Nishikawa method.

Nishikawa didn't live long enough to enjoy much success; he died the next year of stomach cancer at only 35 years of age. Mikimoto recognized that the Mise–Nishikawa method was much more efficient than the one he was using, and in 1916 purchased the rights to it. In honour of Mise, Nishikawa and Mikimoto, the Japanese pearl industry celebrates 1907 as the birth year of the round cultured pearl.[54] In Mikimoto's advertisement that same year, the placid waves rippling from the oyster shell give no indication of the troubled currents beneath the surface.

Mikimoto Shop, Ginza, Tokyo, 1906

The white stone flagship shop on Tokyo's Ginza Street was known as the 'pearl-coloured shop'.[55] Facing a wide, paved boulevard, it represented the highest rung of the real estate ladder. In 1899 Mikimoto had secured a 'starter' shop a few streets back – the retail version of the cheapest house in the most expensive neighbourhood. As business picked up, he moved to a better location in 1902, then in 1906 to the elegant Western neoclassical building on Ginza.[56] Inside, sales staff in formal Western dress assisted customers, many of whom were visitors from overseas.

Before this venture, Mikimoto had difficulty selling his hemispherical pearls as customers misunderstood what they were. He opened his first store not just for convenience or to cut out the middleman but to educate consumers about cultured pearls. He targeted Ginza as it was Tokyo's most fashionable shopping district and because, metaphorically, it faced West. Keenly attuned to the value of his brand, Mikimoto understood that he wasn't just selling pearls; he was selling an idea of modern, technologically advanced Japan to the rest of the world.

Ginza was the fashion heartland of Meiji Japan (1868–1912), a position it still retains. The Meiji Restoration, which succeeded centuries of conservative military rule, ushered in swift reforms designed to turn Japan into a modern industrialized nation. A fire that razed neighbourhoods of traditional wooden houses in 1872 cleared the ground for urban planners to rebuild in stone and brick. Ginza, previously a non-descript merchant quarter, transformed into a showcase of modernity. Its facelift was unabashedly Western. The wide boulevards were inspired by Haussmann's renovation of Paris, although on a more modest scale. Its architecture, in a nod to London's financial district, was designed by the British architects Thomas Waters and Josiah Conder.

In Ginza, Mikimoto's shop rubbed shoulders with businesses that were to grow into powerful corporations, such as the watch company Seiko, the cosmetics company Shiseido and the famed department store Mitsukoshi. The Mikimoto sales staff wore Western jackets and trousers, high starched collars and neckties, and adopted slicked back Western hairstyles. The Meiji government encouraged Western dress, especially for men in public life, and the first adoptees were government officials and the upper classes eager to project sophistication and refinement.[57] Mikimoto's impeccably dressed staff reinforced the image of a progressive, outward-facing, fashion-forward company. For the comfort of overseas visitors unused to bartering, prices were fixed.

One of the stylish advertisements Mikimoto placed in the *Japan Times* from May 1906 touted pearls 'produced by a scientific treatment of living pearl bearing oysters. Tourists are cordially invited to visit and inspect how perfectly those pearls are produced.' Mindful of the popularity of his pearls with foreign buyers, Mikimoto sent his brother, who managed the Ginza shop, and his brother-in-law to the United States in 1904 with samples of his hemispherical pearls. He was building his business during the heyday of aspirational World's Fairs, ideal stages for technological prowess. Mikimoto pearls appeared

One-third-scale model of the Liberty Bell in Independence Hall,
Philadelphia, executed from 12,250 pearls.

in Russia, Brussels, Paris, Milan, St Louis and San Francisco. He
concocted extravagant displays like the model of a famous Buddhist
pagoda made from 12,760 pearls for the 1926 Sesquicentennial
Exposition in Philadelphia. To the delight of patriotic Americans,
he made a pearl replica of George Washington's mansion Mount
Vernon for the 1933 World's Fair in Chicago, which he gifted to the
Smithsonian Institution. The model's lawn, made of 6,000 pearls,
included more pearls than survived in the entire Shōsōin Imperial

THE ILLUSTRATED LONDON NEWS

REGISTERED AS A NEWSPAPER FOR TRANSMISSION IN THE UNITED KINGDOM AND TO CANADA AND NEWFOUNDLAND BY MAGAZINE POST.

SATURDAY, MAY 14, 1921.

The Copyright of all the Editorial Matter, both Engravings and Letterpress, is Strictly Reserved in Great Britain, the Colonies, Europe, and the United States of America.

THE PRODUCTION OF JAPANESE "CULTURE" PEARLS: WOMEN DIVERS ("SEA-GIRLS") SWIMMING TO THE OYSTER-FISHING GROUND.

A sensation has been caused, both in the jewel trade and the world of fashion, by the claim of Mr. Kokichi Mikimoto to have at last perfected the production of "culture" pearls. Other illustrations and notes on the subject appear on later pages in this issue. Here it is sufficient to quote from Mr. Mikimoto's booklet, "The Story of the Pearl," in explanation of the above photograph: "A large part of the submarine work in the oyster-culture of Japan is done by women divers, or 'sea-girls,' as they are called there. This is common in Agu Bay, and in many other localities of the country. There has been a belief from time immemorial that women can work better and stay longer under water than men. The women divers of Ise have often been mentioned in classic literature. These sea-girls are dressed in tight, thin, white garments. Their hair is twisted into a hard knob, and diving glasses are worn. They dive without any apparatus, and stay under water from 60 to 80 seconds at each diving." The floating tubs are used to carry the oysters they bring up.

PHOTOGRAPH BY COURTESY OF MR. KOKICHI MIKIMOTO.

'The production of Japanese "culture" pearls: women divers ("sea girls") swimming to the oyster-fishing ground', *Illustrated London News*, 14 May 1921.

Pearl divers employed by Kōkichi Mikimoto, Japan, 1921.

Guglielmo Marconi visits Kōkichi Mikimoto,
from *Le Vie d'Italia e del Mondo*, April 1936.

Treasury in Nara. In New York in 1939, he exhibited a pearl Liberty Bell, complete with a crack picked out in dark blue pearls.

He expanded his brand overseas, first opening a wholesale venue in London in 1913, followed in 1916 by a retail shop on smart Nanking (Nanjing) Road in Shanghai, close to where department store behemoths Sincere (1917) and Wing On (1918) would soon debut. Retail shops in London, Paris, Bombay, New York, Chicago, San Francisco and Los Angeles would open over the next two decades.[58]

Back in Mie, Mikimoto crafted a tourist experience that never failed to impress overseas guests. An article in *Time* magazine in 1932 noted that 'Pearl King Mikimoto loves to fete Occidental visitors to his pearl farm. First they are given baskets of Mikimoto oysters. Next Mikimoto minions open each guest's oysters, extract the pearls and present them to the guests.'[59] The party would retire to a lunch of fried oysters, a farm-to-fork experience far ahead of its time.

Bergdorf Goodman advertisement
for Mikimoto Pearls, 12 November 1936

A minimalist graphic depicts a white orb glowing against a dark sky. 'Shaped like the full moon. And whiter than the morning star – the divers who got past the sharks might sometimes bring a pearl like that to the surface. Now they grow them in the Japan Sea.' The romantic nature imagery was deliberately unnuanced, glossing over Mikimoto's long struggle to have his pearls recognized as real. They might have looked lunar, but many powerful voices in the gem trade insisted they be labelled artificial. When they were released onto the international market in the 1920s, Mikimoto's biologically engineered pearls raised uncomfortable questions about nature and artifice, authentic and fake, expertise and sleight of hand. Reactions ranged from admiration to disbelief, from anger to panic at what might be unleashed on a market where price cleaved to rarity.

Tiffany's revered gemmologist Charles Frederick Kunz had been sceptical about Mikimoto's early 'button-shaped' pearls. A cross-section, he claimed, revealed a thin, fragile layer of nacre 'deficient in luster'. He snootily pointed out that the awards these pearls had received at exhibitions in Paris, St Petersburg, Tokyo and St Louis were given in the fisheries, not the gem divisions.[60] Tiffany & Co. refused to stock cultured pearls until well into the 1950s.

One of Mikimoto's prized possessions was a silver cup gifted to him by Prince Komatsu Akihito engraved with the words, 'You use the work of man to aid the work of heaven.' It was a Linnean sentiment that perfectly expressed a personal and national agenda to 'husband' natural resources. This entailed raising healthy spat (oyster larvae that have attached to surfaces), nurturing juvenile oysters, stimulating a biological response and eventually reaping a bountiful harvest. Mikimoto's strategy was always to insist that science served nature.

The objections, even when they were framed as scientific or ethical, were mainly economic. In the early twentieth century, natural pearls were a very lucrative business. The pearl fisheries of the Persian Gulf, exploited for millennia, were still the most famous and valuable in the world. Official exports in 1904 were recorded at over £1 million, though by the time they reached the markets of Asia and Europe, the prices multiplied.[61] The Persian Gulf supplied 50 million pearls a year, compared to 20 million from the rest of the world's fisheries. Yet this staggering total was still not enough to meet demand when the United States could snap up 60 million pearls in a single year.[62] The bulk found their way to the great pearl entrepôt of Bombay and from there to the gem markets of the world. High demand inflated prices and supported a robust web of employment. The 80,000 people working the Persian Gulf fisheries were just one part of a network of dealers, syndicates, pearl workers, retailers and speculators.

Mikimoto, who famously said that he hoped to live long enough to see cultured pearls affordable for anyone, threatened to disrupt the delicate balance. The first alarm bells went off in May 1921, when London newspapers reported that cultured pearls had entered the London market and deceived experienced dealers who resold them as natural. Compared to pearls today, these were small in size, at around 2–3 millimetres in diameter. They did, though, compare with small natural pearls. Henry Lyster Jameson, a British zoologist and manager of pearling stations, wrote an influential response for the eminent

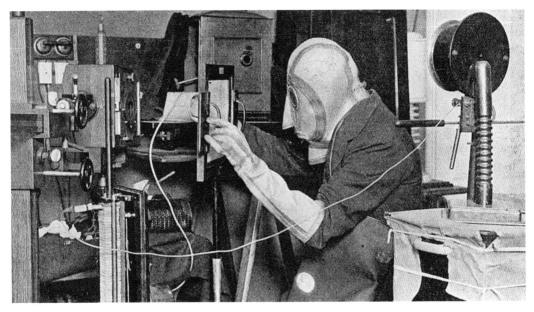

'Taking a radiograph of genuine and cultured pearls', *The Sphere*, 21 May 1921.

Mikimoto sash clip *Yaguruma* (Wheels of Arrows) that can be configured in twelve ways, platinum, 18-carat white gold, cultured Akoya pearls, diamonds, sapphires and emeralds.

scientific journal *Nature*. He reminded readers that Mikimoto had been completely transparent with the public, describing his pearls as artificially produced. As for technique, Jameson referred to Mikimoto's application for a patent in Britain, open for inspection at the Patent Office. He concluded, 'Biologically speaking, the Mikimoto pearl satisfies all the conditions which go to make up a pearl.'[63]

Biology and business were different matters, though. The London Chamber of Commerce denounced cultured pearls as fakes, and in Paris, which had emerged as the centre of the pearl trade in Europe, they were labelled imitations.[64] Mikimoto fought lawsuits throughout the early 1920s but scored a victory in 1924 when a French civil court permitted him to call his gems pearls. In 1926 it was agreed the correct classification was *perles cultivées*: cultured pearls.

An unresolved problem, highlighted by the uproar of 1921, was that cultured pearls were indistinguishable from natural ones; even seasoned gemmologists couldn't tell the difference, which threatened the integrity of the natural pearl market. The only way to know whether a pearl's nacre encased a surgically implanted bead was to slice the gem open. Ironically, it was science, not connoisseurship, that delivered non-destructive methods of testing. During the 1920s, as more cultured pearls entered international markets, pearl testing laboratories sprang up in centres such as Paris, London, Bombay, Amsterdam and New York. A widely used instrument was the endoscope, a hollow needle that used light to detect a cultured pearl's solid core.

That the sole giveaway was internal structure only broadened a cultured pearl's appeal. From the outside, it looked just like the Moon.

Pearl Imperialism

Once they were accepted as real rather than imitation, cultured pearls became a valuable economic asset for Japan. With generous government investment, oyster breeders and pearl farms began to saturate this lucrative market. In 1931 a million pearls were harvested from 51 farms.[65] Quality control, though, was a problem, with some farmers reducing the amount of time between nucleation and harvest, producing pearls with thin, flaky nacre. Recognizing the threat to his brand, Mikimoto staged a spectacular publicity stunt in the city of Kobe, which had emerged as the clearing house for the international export of Japanese cultured pearls. It was, in a sense, a flamboyant launch party for the Japan Cultured Pearl Fisheries Association, a new organization charged with ensuring the high quality of exported pearls. In October 1932, cheered on by a crowd in front of Kobe's Chamber of Commerce, Mikimoto shovelled 720,000 inferior pearls into a brazier. Within minutes, the gems that had spent so little time underwater crumbled to ash.[66]

The spectacle paid dividends; foreign newspapers reported that the price of cultured pearls jumped 30 per cent, and Japan's cultured pearl industry roared onwards. By 1938, 289 farms produced 110 million pearls.[67] The Second World War crippled growth, but robust brands like Mikimoto capitalized on post-war demand in the United States. Periculture was so vital to Japan's economic recovery that it was regulated under a government decree stipulating that pearl-culturing technology should remain a Japanese trade secret. Regardless of their country of origin, all cultured pearls were to be marketed by Japan, and the Japanese Fisheries Agency was to oversee Japanese companies operating overseas. It was a triumph of protectionism that tied up the industry until the mid-1990s.

Post-war Japanese influence on pearl culturing, particularly in marine waters south of Japan, was like the advance and retreat of the tide. As culturing Akoya pearls outside Japan was prohibited, many new ventures focused on *Pinctada maxima* (the gold- or silver-lipped oyster) or *Pinctada margaritifera* (the black-lipped oyster). Mikimoto had experimented with these as far back as 1912 in Japan's southern Ryukyu Islands and later in the Palau Islands in Micronesia, but it was only after the Second World War that pearl culturing outside Japan succeeded on a larger scale. Japanese companies set up operations in Myanmar, which became renowned for high-quality golden pearls. They also established joint ventures in Australia, the Philippines and Indonesia – the warm regions where *P. maxima* thrives. All these enterprises contended with the issues Mikimoto had faced

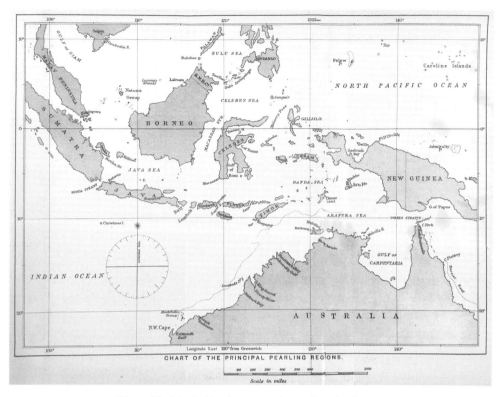

'Chart of the Principal Pearling Regions', Australia and Indonesia,
in Edwin W. Streeter, *Pearls and Pearling Life* (1886).

earlier: technology transfer, securing a workforce, animal welfare, maintaining a healthy aquatic environment and establishing brand recognition. During the following decades, Japanese companies and individual technicians continued to dominate the industry, and it was only with tenacity and grit that Australian entrepreneurs broke free from the hammerlock of Japanese control.

It is fitting that a country big enough to be its own continent is famous for its expansive pearl oyster beds and marble-sized pearls. In the early twentieth century, the pearl oyster grounds off Australia's northern and northwestern coasts extended for almost 4,800 kilometres (3,000 mi.). The South Sea pearls cultured from Australian waters might seem a platonic ideal: round, generously sized, lustrous and bright white.

The history of pearl fishing in Australia, though, is less flaw-less. Before European settlement, Aboriginal peoples had harvested *P. maxima* shells for thousands of years. The oysters that flourish along the north and western coasts not only produce pearls but, more reliably, mother-of-pearl. Unlike the smaller Akoya oysters, the shells of *P. maxima* are as large as salad plates and thick as platters. Historically, pearl shell was the most widely traded cultural object in Australia. Incised in ochre or charcoal with geometric designs, these nacreous 'canvases' were used as gifts and in rituals and rain-making ceremonies.[68] Intimately connected with water, pearl shell symbolized the plenitude of rain and seasonal regeneration. It was, in the words of an Aboriginal elder, 'for everybody – man and woman. This is rain. This everything water.'[69]

Shell was prized far beyond the Australian continent and traded with Indian merchants.[70] The 'pearl frontier', stretching between Australia's northern coasts and maritime Southeast Asia, was a fluid place traversed by ancient trading routes that connected the region with markets in China and beyond. Today's maritime and land borders were largely carved by the Dutch, British and Portuguese, drawn to the Pacific by the quest for spices and tropical commodi-ties.[71] During the period of European expansion, indigenous peoples were dispossessed of territory and resources and sometimes, through necessity or coercion, provided the labour on lands that had once been their traditional homelands or hunting grounds.

For the European colonists, Australia was a blank canvas, a vast, empty place waiting to be shaped and cultivated. For the Aboriginal peoples, the land and ocean – 'saltwater country' – was an animated spiritual geography, constantly remaking itself. For Europeans, oysters were commodities, much like wool, sugar and milk. For Aboriginal people, pearl oysters, with their iridescent shells, were emblems of the Rainbow Serpent, bringer of life, and protector of that most precious commodity of all, water.

Observing the robust trade in pearls and shell along the pearl frontier in southeast Asia, European colonists targeted Australia's oyster-rich northwest Kimberley Coast. Far from metropolitan

FINE PEARL HANDLE AND TRIPLE PLATED TABLE CUTLERY.
Prices Per Set

No. 857. Fruit Set, 6 Nut Picks, 6 Fruit Knives and Nut Cracker, Triple Silver Plated in Fine Kid Case............... $8 40

No. 858. 1 Dozen Extra Quality Pearl Handle Medium Knives, Sterling Silver Ferruled, Triple Plated Blades in Fine Kid Case........... $12 30

No. 859. ½ Dozen Fruit Knives, Solid Steel, Fancy Handles, Triple Silver Plated, put up in Satin Lined Box........... $4 70

No. 860. ½ Dozen Pearl Handle Fruit Knives, German Silver Ferrules, Triple Plated Blades, in Satin Lined Box........... $6 70

No. 861. Carving Set, Extra Large Fine Stag Handles, Sterling Silver Ferrules and Caps, 8½-inch Special French Blade, put up in Fancy Satin Lined Box....................... $12 70

No. 862. Romanesque Sugar Shell and Twist Butter Knife, in Fine Silk Plush Case.............. $2 84

No. 863. Carving Set, 3 Pieces, Knife, Fork and Steel, Fine Stag Handles, 9-inch Spanish Blade....................... $5 20

41

Samples of cutlery with mother-of-pearl handles, featured in
a catalogue for Madson & Buck Company, Chicago, 1898.

F. Nicoud, dessert knife with mother-of-pearl handle, *c.* 1890.

centres and difficult to access, the Kimberley, like Lapland, was considered vacant. But for its Aboriginal caretakers, it vibrated with physical and spiritual energy. Its coastline is a constellation of 2,639 islands, limestone outcrops, inlets and atolls. Underwater, the remains of reef systems nearly 400 million years old harbour prolific marine life. Low tide on the Dampier Peninsula uncovers dinosaur footprints. Each March to May, the sky above the pearling centre of Roebuck Bay shimmers with shorebirds migrating over the East Asian–Australasian Flyway. This coast was a landing stage for Aboriginal peoples migrating from Asia into the new continent.

For European colonists, it was a final tropical frontier, attracting buccaneers, pirates and adventurers. From the 1860s, colonial entrepreneurs developed a lucrative pearl shell industry along the Kimberley and northern coasts. By the end of the nineteenth century, Australia dominated the international market for high-quality mother-of-pearl, with 85 per cent of the crop going to make buttons for handmade clothes and mass-produced ready-to-wear. Pearl shell also made attractive decorative inlay and utensil handles. By 1900 it was Australia's fourth-largest export after gold, timber and wool.

Pearl divers from Torres Strait Islands, standing in boxing poses, *c.* 1920.

Japanese pearl diver, Thursday Island, Queensland, Australia, *c.* 1930.

Pearls themselves, though a welcome bonus, were a by-product of
the pearl shell industry, accounting for only 12 per cent of its value.[72]
    Aboriginal labour and knowledge of oyster beds were vital to
the industry. But being compelled to collect pearl shell – a light and
spirit-filled object – brought despair and darkness to the lives of
indigenous peoples caught up in the drive for profit. In the industry's
early years, colonists forced Aboriginal free-divers to collect shell
under conditions comparable to slavery.[73] Aboriginal women, considered superior divers, were pressed into service.[74] The kidnapping
('blackbirding') and mistreatment of indigenous people were so
egregious that legislation was passed in the 1870s to curb the abuse.
    The colonial 'pearling masters' faced an acute labour shortage.
From the 1870s they began importing indentured workers from

Southeast Asia, a system that persisted into the 1970s.[75] As oyster beds in shallow areas were stripped, pearling masters invested in new technologies such as deep-diving suits with on-deck air compressors installed on large, custom-built pearl luggers – wooden boats 13–19 metres (42–62 ft) long. Japanese labourers, reputedly the most skilled divers, joined multicultural crews of Indonesians, Filipinos, Aboriginal peoples, and South Sea and Torres Strait Islanders. In the town of Broome, which processed 80 per cent of the mother-of-pearl supplied to international markets, the ratio of Asians to white people was fourteen to one.[76] Profit-rich and labour-short, pearling was exempt from the 1901 Immigration Restriction Act, designed to keep Australia white.

Life onboard the luggers was gruelling and often deadly. The boats would spend weeks at sea, contending with cyclones and some of the world's biggest tides. Luggers were crowded, foul-smelling and

Pearling schooner in the Torres Strait, Queensland, Australia, *c*. 1900.

slick with the innards of oysters opened and cleaned on deck. Up to 30 per cent of indentured divers died each year from decompression sickness or drowning.[77] The hundreds of Japanese graves in Broome Cemetery bear poignant witness to the hazards.

Plastic buttons and the Great Depression dented the pearl shell industry in the 1930s, but the Second World War sealed its fate. As 'enemy aliens', Japanese workers were interned, and other migrant workers repatriated. The luggers were requisitioned by the navy or burned to prevent them from falling into enemy hands. The industry was dismantled, but pearling culture, ports, infrastructure, the web of connections across Southeast Asia and, of course, the oyster beds survived and were all vital to the transformation of historical pearling into today's South Sea cultured pearl industry.

To protect the market for natural South Sea pearls, the Australian government banned the production of cultured pearls, effectively leaving Australian biologists dependent on Japanese technology even after the ban was lifted in 1949. The first post-war foray into pearl culturing was a joint venture between the Japanese pearler Tokuichi Kuribayashi and American and Australian partners. Pearls Proprietary Limited opened Australia's first pearl farm in 1956 in Kuri Bay, about 370 kilometres (230 mi.) north of Broome. Named after Kuribayashi, it is an indelible tribute to the contribution of Japanese technology, capital and labour to the Australian South Sea pearl industry. At Kuri Bay, Japanese technicians worked as grafters and adhered to the Japanese policy of not sharing technology with foreigners.[78] They even brought their own nucleus beads with them, having discovered that nuclei made from American freshwater mussel shell produced better pearls.[79]

Pearls Proprietary Ltd was later bought by Nicholas Paspaley, a Greek immigrant who worked in the Australian pearling industry with his family. He purchased his first pearl lugger as a teenager in the early 1930s.[80] After the collapse of the shell industry, Paspaley joined forces with the Japanese firm Arafura Pearling Company and started a pearl farm near Darwin, the capital of Australia's Northern Territory. Again, Japanese grafters provided technical expertise, as they still do today across the industry.[81] Junichi Hamaguchi, an ace technician from Mie, was among the first Japanese to return to Australia in the 1950s and pioneered ways to increase yield. He inserted larger nuclei into oysters and devised a way to double the harvest by implanting a new nucleus into an existing pearl sac.[82] By the 1990s Paspaley emerged as the leading producer of pearls in Australia.

Initially, Australian producers sold all their pearls to Japanese wholesalers, but in Darwin in 1989, Paspaley held the first auction of South Sea pearls outside of Japan. The famous pearl auctions in Kobe were almost ritualistic in their exclusivity. In the immediate post-war period, only about 25 dealers were invited to attend.[83] The prices realized at the Darwin auction surpassed estimates by up to 100 per cent, bringing the Japanese into serious competition with other buyers for the first time.[84] Kobe's famous winter auctions of Akoya pearls remain important, but today major auctions (still invitation-only) are held in Hong Kong and Tahiti.[85]

The Darwin auction was a step, but severing the umbilical cord owed as much to conditions in Japan as it did to Australian prowess. In 1966 Japanese pearl production hit its peak at 230 tons.[86] But this bounty came at a cost. Several thousand pearl farms had sprung up to meet voracious international demand. Oysters were farmed in crowded conditions and, unsurprisingly, suffered waves of mass mortalities. The eminent marine biologist Shohei Shirai sounded the alarm in 1970, arguing that the human and industrial effluent flowing into marine waters threatened the entire industry.[87] His predictions came true in summer 1996, when an epidemic swept through Japan's oyster populations, killing a million oysters a day. By the year's end, 200 million oysters, representing two-thirds of Japan's stock, had died.[88] The blow fell just two years after the devastating Kobe earthquake of January 1995. Large parts of the city, which had so long been the epicentre of the cultured pearl trade, were destroyed, and the all-important auctions disrupted. The pattern of trouble arriving in threes was confirmed in 1997 with the Asian financial crisis. Japanese banks cut lending, and overseas producers lost the Japanese customers who had bought their entire harvests.

Australia responded by focusing on new distribution channels, environmental sustainability and animal welfare. Pearl culturing is a valuable export industry that the government carefully regulates to maintain high quality and prices. A quota system controls the number of oysters seeded. Australia has an environmental advantage as its western and northern coasts are relatively pristine. While other countries with decimated oyster beds have switched to hatchery breeding, Australian producers nucleate wild oysters.[89] This secures biodiversity – a defence against the diseases that devastated Japan's stock in the 1990s, aggravated by reliance on genetically engineered monocultures.[90]

In Australia, pearl farming is capital and labour intensive, with the larger producers owning fleets of ships and aircraft. Technicians nucleate oysters in floating laboratories spotless as Swiss watch factories, reducing fatalities by keeping the animals in their natural habitat.[91] After surgery, the oysters are placed in racks on the ocean bed for several months, and divers turn them to encourage the even growth of the pearl sac.[92] They are X-rayed back onboard ship to ensure that they haven't rejected their nuclei, then are suspended from submerged longlines – ropes 100 to 200 metres (330–660 ft) in length anchored between buoys. While the traditional Japanese method of hanging racks from floating rafts encouraged over-crowding, the longline system gives the animals space and promotes healthier populations. This attention to the environment and welfare of animals may seem progressive but harks back to an indigenous tradition of stewardship that is thousands of years old.

CULTIVATING BRANDS

The success of today's South Sea pearl industry owes a lot to the lustre of the brand. South Sea pearls account for over half of the *value* of global production of saltwater pearls yet are a tiny fraction of the pearls on the market. By volume, Chinese freshwater pearls have captured a whopping 98 per cent market share.[93] South Sea pearls, pristine Akoyas and Tahitian black pearls must compete not just with each other but with these increasingly choice freshwater specimens.

Anil Maloo, necklace, bracelet and earrings of cultured white
South Sea pearls, produced by *Pinctada maxima* oysters.

In the 1960s Chinese growers developed a nucleus-free culturing technique, implanting mantle tissue in *Cristaria plicata* mussels. At first, the yields were clumsy 'rice krispie' pearls, resembling crinkly kernels of breakfast cereal. When farms switched to *Hyriopsis cumingii*, the triangle mussel, during the 1990s, shapes and sizes improved. Freshwater pearl farming is now a booming industry in Zhejiang and Jiangsu Provinces, where farmers have converted rice paddies into shallow pools that shimmer across the flat Lower Yangtze plain. China's spectacular production rests on the triangle mussel's high tolerance for water pollution (marine oysters demand pristine environments), as well as its ability to accommodate thirty to fifty pearls at a time. Treatments such as bleaching and dying (more commonly used in China) can transform dross into gold. Chinese freshwater pearls now flood markets in such quantities that they are commodities rather than luxuries, considered by some in the Akoya old guard as 'junk'.[94] Under this onslaught, South Sea, Tahitian and Akoya producers have latched on to the notion of luxury to weave narratives justifying these pearls' considerable higher prices.

The value of a glossy, cherry-sized South Sea pearl as an aesthetic object lies in physical qualities that can, to an extent, be objectively measured. But its worth as a status symbol rests on the intangible magic of the brand. Historically, the most successful pearl-producing regions and companies braid stories of the environment, technological prowess and celebrity allure. Linnaeus recognized the power of branding when he tried to tempt the Swedish government to buy his pearl-culturing technology in the 1760s. He stressed that cultured pearls would be a Swedish invention, a brand dependent on Swedish rivers and home-grown scientific achievement that would boost the national economy.

Mikimoto was preternaturally aware of the value of his brand, building a company whose watertight reputation persists to this day. His legendary showmanship helped – most famously his cremation of nearly a million inferior pearls. But so too did his calculated courting of celebrity. He secured the patronage of the Japanese imperial family, gifted pearls to British royals and welcomed American politicians to Pearl Island. When baseball player Joe DiMaggio gifted Marilyn Monroe a Mikimoto necklace during their honeymoon in Japan, the company basked in the reflected glamour.[95] In homage to Princess Grace of Monaco, Mikimoto released a Princess Grace Collection in 2001, with advertisements conflating the luminous beauty of the princess and the pearls.

Today, celebrities wearing Mikimoto pearls flit across the company's social media channels, but courting the avant-garde is another tactic from Kōkichi Mikimoto's playbook. Ginza 2, the new Tokyo store designed by Toyo Ito in 2005, appeals to younger consumers.

With a lustrous pink steel skin, punctuated by windows like efferves-cent bubbles, it is a polished jewel box of a building, as aspirational as Mikimoto's first European-style retail space in Ginza. A century after Mikimoto's round cultured pearls disrupted the market, it's easy to lose sight of how revolutionary these were. It's fitting, then, that another radical Japanese inventor, Rei Kawakubo, is turning pearls' associations with traditional feminine elegance upside down. Her company, Comme des Garçons, has designed a line of pearl necklaces for Mikimoto that combine the clean white lustre of Ayoka pearls with gleaming silver chain. Like Kōkichi Mikimoto, Kawakubo has spent her career challenging the status quo and blurring distinctions between East and West, Japanese and international, art and nature. Her line of gender-neutral necklaces is a fresh and timely artistic intervention and a reminder that, historically, men too have enjoyed flaunting pearls.

Since the 1970s, Mikimoto has focused on luxury jewellery, a direction forged earlier when Kōkichi Mikimoto founded a metal-working factory in 1907 to create the settings for his pearls. Mikimoto no longer farms its own pearls – water pollution remains a problem, and Japan's Akoya oysters have faced further mass die-offs – so an emphasis on high design is a sage strategy.

In contrast, the direction taken by high-end producers of South Sea pearls has been to market the unique characteristics of the en-vironments that nurture the pearls. In 1964 the Duchess of Windsor purchased a necklace of 29 cultured Australian pearls from the French firm Van Cleef & Arpels, lending celebrity cachet to these unfamiliar gems, but their popularity in the United States owes much to the influence of uber-dealer Salvador Assael. After the war, Assael traded Swiss Army watches to the cash-strapped Japanese for pearls. He soon turned his attention from dainty Akoyas to the larger South Sea pearls that Japanese growers were farming off Australia. Knowing these marble-sized gems would appeal to Americans, he began selling to major jewellers such as Harry Winston and Van Cleef & Arpels. In 1994 Paspaley and Assael helped launch the South Sea Pearl Consortium to promote these pearls, and two years later, they formed a joint venture to market them in the United States.[96]

Necklaces of superb cultured South Sea pearls can realize several million dollars, a price tag firmly tethered to their place of origin – or more precisely, a highly curated idea of their place of origin. They are marketed as the artful outcome of their pristine environments and clean technology. They embody place – Antipodean light shimmering through the water column – as well as sustainability. They present pearl farming as a coalition between human and nature that ensures the well-being of the ocean, its coastlines, and human and animal communities.[97]

A pearl is an oyster's autobiography, but it is also a carrier of tales about our own values and aspirations. We use pearls to fashion ourselves and to speak of our status and environmental awareness. We culture pearls, and pearls culture us.

❧

FOUR

The Seven Pearly Sins

Pearls' meanings, like nacre, are flexible. Nacre is an elastic brick wall consisting of brittle, inorganic aragonite platelets in a soft, organic matrix. The duality built into the very microstructure of the pearl is echoed in its symbolism. Cultured pearls are both affordable and exclusive. *Pearl* is the title of a Victorian pornographic magazine and an elevated medieval English meditation on purity.[1] A pearl is a Scheherazade, spinning tales about our love of beauty and goodness, but also our envy and pride. Like many precious objects, pearls bring out our worst, avaricious, deceitful dispositions, but also our better angels. They can embody all the deadly sins, but just as easily, their opposite virtues.

LUST: AMOROUS *AMAS*

In a scene from *Tampopo* (1985), the Japanese film director Juzo Itami's satirical paean to foodie pleasure, a white-suited gangster watches a young female free-diver (*ama*) emerge from the waves with a basket of oysters. She opens one for him with her knife, and he nicks his lip on the sharp shell. She shucks it for him, and he sucks it from her palm, leaving a bright drop of blood in her hand. She pulls him into a ravenous, mollusky kiss. Six of her fellow divers watch from the shallows. It's a riff on the pleasures of food and flesh – and on the erotic charge of the white-costumed *ama*, incorrectly but persistently framed as a pearl diver.[2]

Thirty years later, *amas* in Mie, long home to Japan's cultured pearl industry, forced the city of Shima to abandon its new manga-inspired mascot of a sexy young *ama*. Professional *amas* decried it as child pornography. The 'cute and adorable' mascot was seventeen years old, in the market for a boyfriend, and dressed in a pearl-buttoned outfit that barely constrained her perky, plump breasts.[3] The real *amas* were grandmothers in their sixties. A rare score for reality in an otherwise boundless ocean of lust.

Attitudes to divers and pearls weren't always so lascivious. Some references to the gems in the eighth-century *Manyoshu* (Collection of 10,000 Leaves), the oldest extant anthology of Japanese poetry, are positively chaste:

> Of those abalone pearls
> That Suzu's fisher-maids dive for,
> Crossing over, I hear,
> To the holy isle of the sea,
> Would I had many – even five hundred!
> To my dear loving wife,
> Who ever since we parted sleeves,
> Must be sighing after me,
> Counting the weary days and months
> Passing the night in a half-empty bed.[4]

So what was the journey from the uxorious to the salacious? How do we move from a half-empty bed to two octopi performing cunnilingus on an enraptured young diver? What makes a woman turn from sighing to writhing? Over the years, reactions to Hokusai's *Diver and Two Octopi* have ranged from horror to fright to outrage at the diver's violation. Down this road lies tentacle porn: bestial, degrading, perverse, probably criminal.

Hokusai's viewers would have laughed, both at the image and the ways it has since been misconstrued. Although now endlessly reproduced as a titillating curiosity, it originally belonged in a three-volume album of erotica, *Kinoe no komatsu* (Young Pines of Kinoe), published in 1814. Each volume begins with a portrait of a beautiful woman, progresses through shots of couples engaged in breathless intercourse and climaxes in a vaginal close-up. The text surrounding *Diver and Two Octopi* includes a conversation between octopi and diver in which she makes her ecstasy quite explicit: her pleasure is eightfold.[5]

The text also alludes to an ancient folk tale, providing more context for this image and the development of the sexy diver trope.[6] In this tale, a female diver confronts the Dragon King of the Sea to retrieve a precious gem – in some tellings a pearl – that he stole from her family. Pursued by an underwater army of vengeful octopi and other sea creatures, she slits open her breast, conceals the gem

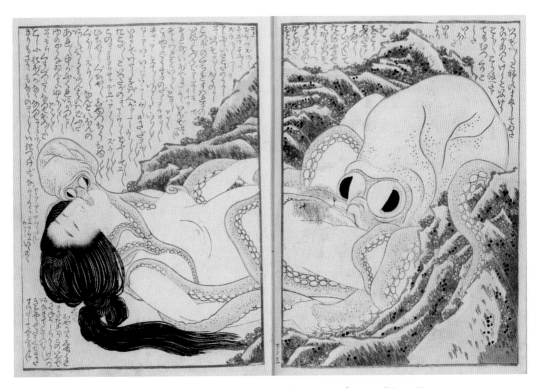

Katsushika Hokusai, 'Diver and Two Octopi', in *Kinoe no komatsu* (Young Pines of Kinoe), 1814, woodblock print; ink, colour and metallic pigment on paper.

Kitagawa Utamaro, *Abalone Divers*, c. 1797/8, colour woodblock print.

and evades capture. Hokusai turns this tale of wifely virtue and self-sacrifice on its head. Both octopi and diver are too engrossed in their pleasure to attend to other business.

This most lustful of images is a satire, fuelled by a culture that had long connected women, water and sex.[7] Water is the feminine *yin* element: dark, hidden, moist, flowing. The underwater world is female. Men might navigate the sea in boats, they might be pilots, sailors, fishermen and merchants, but only women have intimate knowledge of the ocean floor. Women know what seasons to gather seaweed or Akoya oysters, where the precious abalone feed, when to harvest and when to conserve resources for the future.

One of Japan's foremost post-war writers, Yukio Mishima, based his novel *The Sound of Waves* on his experiences in the traditional fishing community of Kamishima, in Mie Prefecture. The mother of the main character, a highly-skilled *ama*, acknowledges that while men might go sailing,

Women, not destined for that wide world, cook rice,
draw water, gather seaweed, and when summer comes
dive into the water, down to the sea's deep bottom. Even
for a mother who was a veteran among diving women this
twilight world of the sea's bottom was the world of women . . .
All this she knew. The interior of a house dark even at noon,
the somber pangs of childbirth, the gloom at the bottom
of the sea – these were the series of interrelated worlds in
which she lived her life.[8]

The ocean is a womb that births the valuable sea creatures that a skilled *ama* can harvest and sell for a comfortable profit. Seashells have clefts like vulvas, and the labial mantles of oysters enfold soft, glistening flesh. Men fish for bonito. Women fish for sex.

And they fish naked. For centuries Japan's *amas* wore little but a loincloth. Their nakedness and income-earning abilities lent them a reputation for financial and sexual freedom. Their reputed services might even extend to rescuing shipwrecked sailors and reviving them against their naked breasts.[9]

In Japanese fishing villages, however, the naked breasts of hardworking *amas* weren't sexualized. But by artists they were. In the mid-twentieth century, Japanese photographer Yoshiyuki Iwase published images of topless and nude *amas* in his home prefecture of Chiba. Some documented a disappearing way of life. Some blur the boundaries between ethnography and eroticism. Some are unabashed celebrations of his models' bodies, posed naked on rocks, in sand, or entangled in fishing nets. But it was the Italian photographer and anthropologist Fosco Maraini who introduced the *ama* to a wide overseas audience. His book *The Island of the Fisherwomen* is a work of lush romanticism. Translated into English in 1962, his breathless prose and images of *amas* taken during the summer diving season on Hekura island celebrate their 'primitive vitality' and 'primordial humanity'.[10] The photographs linger on nubile breasts and glistening skin. The mirage of the *ama* was so enticing that it inspired Ian Fleming's James Bond novel *You Only Live Twice* (1964) and its film adaptation in which Sean Connery and a semi-clad 'Kissy Suzuki' frolic through a rocket base disguised as a volcano.[11]

In popular and literary culture, there was an unabashed fixation on *amas'* breasts. *The Sound of Waves* is a relatively short novel in which Mishima devotes a generous word count to breasts:

> All of their breasts were well-tanned, and if they lacked the quality of mysterious whiteness, still less did they have the transparent skin that reveals a tracery of veins. Judging merely by the skin, there seemed to be no particular indication of any sensitivity. But beneath the sun-burned skin the sun had created a lustrous, semi-transparent color like that of honey. The dark areolas of the nipples did not stand out as isolated spots of black, moist mystery, but instead shaded off gradually into this honey color.[12]

The *amas'* breasts were a boon but eventually a national embarrassment. Kōkichi Mikimoto relied on the services of Mie peninsula *amas* for many of the tasks around his pearl-culturing farms. Historically, these divers gathered seafood, not pearls, which were an occasional, though welcome, find. When Mikimoto set up in Toba, he relied on *amas'* intimate knowledge of edible molluscs and the ocean. They collected juvenile oysters and moved them around within the shallows to ensure their growth. They retrieved them for surgery and returned them to the water. When weather or red tide threatened, the agile *amas* could swiftly move the rafts to safer locations. They became essential to Mikimoto's operation and a major tourist attraction. They were the best-known *amas* in Japan and had become intimately associated with pearl-diving. They were industry workers who also performed for tourists, dismissed by the myth-loving Fosco Maraini as 'not the real thing'.[13]

At first, Mikimoto's *amas* worked bare-breasted. As his enterprise began to attract overseas visitors and trade delegations, he devised a modest white cotton uniform based on the clothing the divers wore in their huts after work. This is the sedate costume, not

Amas at Mikimoto Pearl Island, Toba, preparing a demonstration dive.

Ama diving demonstration, Mikimoto Pearl Island, Toba.

unlike a full-coverage 1920s bathing suit with a snug hood, still worn by *amas* today on Mikimoto Pearl Island for diving demonstrations. Brochures for tourists describe it as a 'traditional diving outfit'.[14] It's also the outfit worn by Juzo Itami's amorous young diver in *Tampopo*.

And so this, by way of centuries of watery association, metaphor and misconception, is the route to the city of Shima's hapless sexy *ama* mascot. And this is how the ocean floor became a bed.

PRIDE: PEARLS FOR ISIS

The scene has been painted so many times. A fair-skinned Queen Cleopatra beguiles the handsome young Roman general, Mark Antony, over dinner. In her palace in the royal city of Alexandria, everything gleams: the marble statues, the silver goblets, the gilded platters, the heaped satins, the flushed skin of servants staggering to the table with tureens of stew made from fish caught early that morning on the banks of the Nile. And brightest of all is the pearl, as big as a duck's egg, which Cleopatra has just plucked from her ear and now suspends over her wine glass. Mark Antony looks puzzled. Or maybe he's just dyspeptic. He has, after all, been gorging for days on wild boar. The palace cook has been roasting eight at a time for a dozen debauched diners.[15] They called themselves the Inimitable Livers, though judging by the quantities of wine chilling by their feet, the Ravaged Livers would also seem apt.

We know what's coming, thanks to the Roman polymath Pliny the Elder, who told the story in his *Natural History*. Like gossip repeated too many times, the details are hazy, but we know Cleopatra's pearl earrings are the largest in history, handed down, Pharaoh to Pharaoh, to this last decadent queen of the Nile. We know that this meal, however elegantly arrayed on lace-edged linens still creased from the iron, is, in fact, an extreme eating contest, pitting Egyptian against Roman to decide who could consume the most expensive meal. And we know that Cleopatra is just about to dissolve her enormous pearl in her goblet, and in one gulp, swallow the cost of a fashionable Mediterranean villa.[16] The extravagance of the gesture gave priggish

Romans plenty of ammunition against the wastrel queen. Especially as Mark Antony's long-suffering wife Fulvia, left behind in Rome, was dutifully raising troops to support him. Cleopatra was a seductress, a whore, insatiable in bed, a sexual deviant. Her deadly sin, according to her grey-faced Roman critics, was lust. Or greed. Or avarice.

In fact, it was pride.

Cleopatra VII had every reason to be proud. She had survived the machinations and murder that laced Ptolemaic life not just during her reign but for the three hundred years of the dynasty's existence. At fourteen, she shared the throne with her father. Her older sister was beheaded for insurgency. Cleopatra may have watched the execution. At eighteen, she became Pharaoh with her adolescent brother. Cannily aware of the power of pageantry, she accompanied the newly chosen sacred Buchis bull nearly 1,000 kilometres (600 mi.) up the Nile towards Thebes, then rowed him across the river for his dedication.

Giambattista Tiepolo, *The Banquet of Cleopatra*, 1744, oil on canvas.

The Ptolemies weren't native Egyptians but Macedonian Greeks; as the last Ptolemaic Pharaoh, Cleopatra understood the importance of paying tribute to Egyptian gods.

Three years later, she was locked in a toxic rivalry with her 'loving sibling', who was also her husband. He expelled her from Egypt, and she fled to Syria to plot her revenge. Barely in her twenties, she raised an army of mercenaries and marched back through the arid Sinai to Egypt to challenge her brother/spouse. Within months the boy king was dead, drowned in the Nile.[17] His co-conspirator, Cleopatra's perfidious younger sister Arsinoe, was paraded in chains then exiled to the Temple of Artemis in Ephesus, in modern-day Turkey. Cleopatra later had her sister slaughtered on the temple steps – a stunning defilement of a sacred space.[18]

Cleopatra might have been ruthless, but she was also charismatic.[19] Plutarch has her smuggled into Julius Caesar's quarters in Alexandria in a bundle of bedsheets (Hollywood ramped up the spectacle by delivering Elizabeth Taylor in a crimson Persian rug). It might have been undignified, but Caesar was captivated. In seducing Caesar, she dazzled a middle-aged man tired of love. She teased him with breasts wholly visible beneath gauze, conquered him on couches studded with gems. She painted her face with Egyptian cosmetics and decked herself out in pearls from the Red Sea.[20] The outcome, in 47 BCE, was a child, little Ptolemy Caesar, or Caesarion. To celebrate his birth, Cleopatra had coins struck depicting herself as the goddess Isis nursing the infant Horus. It wasn't the first time that she claimed to be a deity – she had been divine from the start. The Ptolemies weren't just intermediaries between the Egyptian gods and their people; they *were* gods.[21]

Wishing to be God, striving for the highest seat of heaven – in the Judaeo-Christian tradition, this is the pride, the unforgivable hubris, that got Lucifer hurled out of heaven. But the Ptolemies legitimized their authority by identifying with Egyptian deities. Cleopatra very publicly exploited her role as Isis, one of Egypt's most revered deities. Isis's tears caused the Nile to swell and flood, inundating the soil with water and the mineral-rich silt that made the land fertile. Without

Isis, there could be no barley or wheat, figs, pomegranates or grapes, no orchards or gardens, no honey or palm wine, no papyrus. There would be no plump waterfowl to roast at banquets, no onions, leeks or garlic to toss with pigeon, no lotus to perfume the air.[22] Isis separated Earth from the heavens and commanded the Sun, Moon, tides and sea. In this nature-loving culture, she cradled life itself.

There were temples to Isis throughout Alexandria and more modest shrines in people's homes containing small terracotta statues of the goddess. She was offered fruits and flowers, and when someone wished to make a lavish gift, pearls.[23] She was worshipped across Egypt and far into the Roman world. Cleopatra wasn't the first Egyptian queen to identify herself with the goddess, but she was Isis's most successful human incarnation.[24] When Elizabeth Taylor's Cleopatra announces to Caesar, 'I am Isis. I am worshipped by millions who believe it,' this wasn't Hollywood hyperbole.

And in this first-century Mediterranean melting pot of cultures, where deities did double duty in Egyptian and Graeco-Roman worlds, Cleopatra also became Venus Aphrodite. During her visit to Rome in 46 BCE, a smitten Caesar installed a life-sized golden statue of her in his family temple. Dedicated to Venus Genetrix, the Mother Venus, from whom he claimed ancestry, the temple conflated the two goddesses in a stunning (and censured) display of adulation.[25] He also offered a piece of armour made of pearls in thanks to the goddess for his numerous overseas conquests. It was the perfect tribute to Venus, born from the sea and long associated with pearls.

Cleopatra was again in Rome when Caesar was assassinated in 44 BCE. It was yet another bloody coup she survived. Three years later, she was revelling with Caesar's next-in-command, Mark Antony, who was in Egypt to raise money and support for Rome's eastward expansion. Their union was celebrated as the meeting of two gods, Dionysus and Aphrodite.[26] They held lavish dinner parties and made their cooks prepare the same meal multiple times, so that if they lingered too long, the next course would arrive without a gobbet of congealed fat or lukewarm sauce. They drank to excess, played games of dice, hunted and enjoyed each other in every way. Like A-list celebrities, they delighted in dressing up, melting incognito into a crowd at night. Their tastes were a little warped; she liked to dress as a maid and Antony as a slave. He reportedly enjoyed the beatings.[27]

They adored games, which explains the infamous wager about who could host the most lavish banquet. Most often, artists have played on the pair's fabled oenophilia by depicting Cleopatra about to drop her pearl into a goblet of wine. Given the Egyptian nobility's appreciation of wine and their skill in viniculture, it seems a fair assumption. But Pliny is quite clear that what her servants placed on the table was

> a single cup of vinegar, the sharpness and power of which disintegrates pearls to a pulp. She was wearing in her ears that especially unusual and truly unique work of nature. And so, with Antony eagerly anticipating what she would do, she took one off and dropped it in, and when it was wasted away she swallowed it.[28]

The artists improved on Pliny. What wine lover would drink vinegar when she could have the finest grape wines or even the Egyptian beer that was almost as good? How would swigging sharp vinegar impress Dionysus, god of wine and ecstasy? If the point of the fable was to point out Cleopatra's extravagance, wine was a much better medium.

But the chemical reaction requires vinegar. Pearls are primarily calcium carbonate, which reacts with the acetic acid in vinegar. What Pliny doesn't clarify is how long it would take for the pearl to dissolve. In paintings of Cleopatra's banquet, we feel poised on the dramatic moment, watching the pearl about to splash into the ruby-red wine. We anticipate the fizz of carbon dioxide as the reaction gets underway, turning the wine into a chalky pink antacid. Perfect medicine for ravaged livers.

Except wine doesn't dissolve pearls. The right solution of vinegar, though, does. It might take 24 hours or more for the reaction to complete, leaving a small sludge of protein gel at the bottom of the glass. Not as delicious as the best Egyptian wines, but palatable enough.[29]

Critics used the episode to prove Cleopatra's wastefulness, but there's more to the story than this. Pliny tells us that she swallowed her pearl out of pride to prove her superiority to Mark Antony. She was proud of Egypt's wealth. Her country was both gold mine and granary and encompassed the Red Sea pearl fisheries that supplied the gem to a pearl-hungry empire.[30] But more importantly, she was proud of her accomplishments. 'When Antony was fattening himself every day at decadent banquets', Cleopatra was devising a clever scientific spectacle.[31] She was sharp-witted, he an artery-clogged, slow-moving target. Cleopatra had been educated in Alexandria, the intellectual centre of the Mediterranean world, in a culture that offered women many more freedoms and rights than in Rome. She had received the best Greek education available and benefited from the minds of the best scholars the Ptolemies brought to court. She had access to the Great Library of Alexandria and its museum. She spoke several languages, including Egyptian, and studied mathematics, philosophy, music and astronomy.

Her detractors were to claim that she experimented with poisons, with which she tortured prisoners. More likely, she was interested in botany, medicine and the sciences.[32] Egyptian culture was dedicated to understanding the chemical properties of plants and materials to heal the living and mummify the dead. She would have known, for instance, that the salt natron, which contains sodium carbonate and bicarbonate, could be used to mummify a body. Possessing an education and intelligence like this, she doubtless knew that a pearl would dissolve in acid. Like Isis, she controlled nature. And in controlling nature, she became godlike.

In the end, it was pride and possibly poison that killed her. With Mark Antony dead and Alexandria under attack from Roman forces, Cleopatra committed suicide rather than face capture. Maybe she remembered back, half a lifetime ago, to her humiliated sister Arsinoe, dragged through Rome in chains. Plutarch claims she had an asp smuggled into her rooms in a basket of figs and induced it to bite her arm, but more likely, she swallowed poison, efficient to the last.

94 Unknown artist, *Head of Venus*, Roman, *c.* 100 CE, bronze with gold and pearl.

With the collapse of the Ptolemaic Dynasty, Egypt's pearls flowed westwards into Roman coffers. The largest might have been the pearl Cleopatra didn't dissolve, the one in her other ear the day of the infamous banquet. It reappeared five years after her death, cut in two, decorating the ears of a statue of Venus in the newly dedicated Pantheon in Rome.[33] A pearl returned to Isis.

WRATH: 'A LION UNCONTROLLABLE'

That Genghis Khan was wrathful is born out by the numbers. In his quest to forge a new Mongol world order, he razed cities, culled armies, terrorized civilians and laid waste to civilizations. His 1219–21 campaign to conquer the lands of the central Asian empire of Khwarizm, ruled by the Turkic sultan Muhammad II, was especially devastating. Persian chronicles document the slaughter of 1,747,000 at Nishapur in today's Iran. The thirteenth-century Persian historian Juzjani claimed that 2,400,000 were killed at Herat.[34] Over the centuries, the numbers, like the adjectives used to describe the rapacious, savage, tyrannical Khan, multiplied. Today he is widely credited in popular media with the murder of 40 million people, or more than 10 per cent of his world's population.

After his death, what remained was his black spirit banner, fashioned from stallions' hair attached to the shaft of a spear. Once there were two spirit banners: a white one for peacetime and a black one for war. The white one disappeared, but Genghis Khan's black spirit banner, safeguarded for centuries in Mongolia, was believed to embody his soul.[35] It was wrath that survived.

Material objects like the spirit banner, with tail hairs from his best stallions, flowing in the wind beneath a glinting blade, were at the heart of Genghis Khan's empire. The creation of the vast Mongol Empire, which at its height stretched over 28.5 million contiguous square kilometres (11 million sq. mi.) from Korea to Hungary, and Siberia to the Persian Gulf, owed as much to objects as to wrath.[36] Sometimes the two collided.

Two objects provoked Genghis Khan's first act of murder. The story is told in *The Secret History*, a Mongol book of epic poetry and narrative completed shortly after Genghis Khan's death. After the assassination of his father, the child Temujin, who years later would become the warrior Genghis Khan, fished and hunted with his three brothers to help his family survive. An argument flared over a small fish and a lark that Temujin believed his half-brothers Begter and Belgutei had stolen from him. 'They have snatched something . . . How can we live together with them?' he protested to his mother when she scolded them for fighting over trivialities. In retribution, he and his younger brother ambushed Begter while he was watching over the family's horses. For the theft of two small objects, the boys shot their brother with arrows and left him to die on a small hill overlooking the steppe. Their furious mother likened Temujin to 'a lion uncontrollable in its rage'.[37]

Decades later, during the Mongol invasion of Khwarizm, it seemed that wrath motivated another murder over a small object. The Persian court official and historian Ata-Malik Juvaini documented the Mongol army's 1221 conquest of the city of Termez in present-day Uzbekistan. The inhabitants' confidence in their troops, weapons and sturdy city walls rising above the Oxus River encouraged their spirited resistance. After eleven days of fighting, though, the city fell, and its people were herded onto the plain and butchered. The soldiers found one woman still alive who begged them, 'Spare my life and I will give you a great pearl which I have.' When her captors understood that she had swallowed it to preserve her life, they ripped open her belly and recovered a small trove of pearls. Suspecting that others had followed her example, Genghis Khan ordered his soldiers to eviscerate the dead in search of swallowed pearls.[38]

The story is an example of Mongol efficiency during the conquest of Central Asia, in slaughtering the population of one city to dissuade resistance in the next. Its narrator, Juvaini, had been born the year after Genghis Khan's death and served as an official for the Khan's grandson Hulegu. As an ally of the Mongol court, Juvaini became one of its most important chroniclers, drawing on his own experience and Mongol sources. This small detail about pearls is recorded decades after the fall of Termez. At face value, even for a chronicle of conquest, it seems gratuitously gory as well as puzzling. By this date, pearls flowed into Mongolia through tribute, plunder and the control of trade routes. To recover a few pearls from the intestines of fallen enemies seems almost parsimonious. The tale does, though, affirm the value of pearls to Genghis Khan and the Mongol nation.

Throughout their campaigns, the Mongols looted saltwater pearls from enemy cities and demanded them as tribute from conquered peoples. But Mongolians had an earlier and more intimate experience of pearls. *The Secret History* was composed in the Mongolian language, and although the author is unknown, it was written by someone very close to Genghis Khan. It starts with an origin myth: the story of a blue-grey wolf who settled with a deer at the head of the Onon River. From its source in the Khentii Mountains on

the Mongolian Plateau, the Onon runs through Mongolia, looping
through forests and marshlands. Its cold blue waters flow beneath
craggy outcroppings of rock and are home to trout and giant Siberian
taimen fish. It runs through Mongolia for 300 kilometres (185 mi.),
joining other waterways to meet up with the Amur, or in Chinese,
the Heilongjiang, Black Dragon River, which empties into the Pacific.
Together, these rivers form one of the world's ten longest waterways.
Palaeontological records reveal that a species of freshwater pearl mussel, *Margaritifera dahurica*, likely inhabited the upstream section of
the Onon since the Late Miocene, about 6 to 7 million years ago. The
pearl mussels have certainly been there, in the Mongolian homeland,
for the past 100,000 years.[39]

The Secret History relates that the child Temujin was born beside
the Onon, where his ancestors had hunted with falcons. His clan lived
on the border between the open steppe and the forested mountains
to the north. He was born to the sound of rushing water, and water
flowed through every episode of his childhood. When the Onon froze
over in winter, he played on the ice with boyhood friends. And when
his father died and his clan abandoned the family, his enterprising
mother hitched up her skirt and plunged into the Onon to forage
for food. She sustained her family with the provisions of the river
country: wild leeks, onions, garlic, lily bulbs and crab apples. As soon
as they were old enough, Temujin and his brothers fished 'Mother
Onon' for salmon and fingerlings. The river sustained life, and later it
returned Temujin to his kin; as a youth, he escaped kidnappers first
by floating in the river, then tracking upstream to rejoin his family.

Long associated with the vast expanses of the steppe, Genghis
Khan was equally at home with water. The animism that infused
steppe culture taught him that the essence of life was water, wind and
light.[40] Throughout his life, he revered water as a source of spirit. In
1206 Temujin became Genghis Khan by the headwaters of the Onon
beside the sacred mountain Burkhan Khaldun, in a ceremony that
marked the unification of a million nomadic people into the Great
Mongol Nation.[41] Over the following years, Siberian tribes to the
north and Uighur peoples who farmed and traded along the desert

oases to the west joined the new nation. A condition of the alliance was the payment of tribute. The north supplied sable fur and hunting birds; the Uighurs delivered gems and pearls.[42]

During the next decades, the Mongolians amassed enormous quantities of pearls. They had access to freshwater pearls in the Onon and the waterways of the Amur River basin. But they procured massive stores of large saltwater pearls through diplomacy and plunder. South of the Gobi, the wealthy cities of Jin-dynasty China were an irresistible target. Genghis Khan's army crossed the desert in 1211 and tracked across northern China, systematically looting rich urban centres. When the capital city Zhongdu (modern Beijing) finally fell in 1215, Mongol slaughter left the ground around the city slick 'with human fat'.[43] The city's fabled wealth – silks, lacquered furniture, porcelains, perfumes, cakes of tea, medicines, gems and pearls – was loaded onto carts for the journey back to the Mongol homeland. There was too much booty for the camels, staggering across the sands. Enslaved peoples joined the caravans northwards, leaving Zhongdu to the piles of bones witnessed by appalled ambassadors from Central Asia sent to survey the damage. Some of the pearls the Mongols looted from Zhongdu had been stolen one hundred years earlier from the previous rulers' burial vaults – wrath recycled.

After China, Genghis Khan's army set out in the winter of 1219–20 to conquer Central Asia, laying waste to the fabled cities of Bukhara, Samarkand, Peshawar, Tabriz and Tbilisi. The spoils, including quantities of large Persian Gulf pearls, were redistributed among the Khan's followers and released back into the economy. The Mongol empire was awash with pearls and luxury goods. Mongolians became rich, enslaving captives for manual labour. They settled down as consumers, addicted to the status symbols of silk and pearl-embroidered clothes. The most sought-after commodities were horses, with silk and pearls not far behind. The Mongol elite distributed pearls by the scoopful; in 1223 Genghis Khan gave a favoured commander a pitcher of them.[44]

They had no material need, then, to rummage through the intestines of murdered rebels for pearls. But when the citizens of Termez swallowed their pearls rather than relinquish them, they were doing something more than denying an enemy army their possessions. Again and again, the destruction wreaked by the Mongol army is described not as pillage but as righteous wrath. Juvaini tells us that after destroying Bukhara, Genghis Khan told its surviving leaders, 'I am the punishment of God. If you had not committed great sins, God would not have sent a punishment like me upon you.'[45] He was the Whip of God.

Mongols believed that Genghis Khan had been sanctified to restore order to an unruly world. The wrath that rained onto cities like Termez was heaven's wrath, and denying pearls to Genghis Khan was denying the will of God. A pearly metaphor relating to one of the Mongols' mythical ancestors makes this even more explicit. Alan Qo'a is a mother figure who appears in *The Secret History* and in an account by Rashid al-Din, Persian statesman and author of the world's first world history, *Collector of Chronicles*. For Rashid al-Din, the 'pure womb' of Alan Qo'a is the shell that nurtures Genghis Khan, 'the incomparable pearl'. All future emperors would descend from Genghis Khan, like pearls growing inside a shell.[46]

The ill-fated captives in Termez were not just stealing pearls from the Mongol army; in Mongol eyes, they were challenging the God-given right of Genghis Khan and his descendants to rule the world.

GLUTTONY: APPETITES INDULGED

His parents dragged him on military campaigns and dressed him in a child's version of an army uniform, complete down to the *caligae*, the signature hob-nailed sandal-boots. They called the adorable miniature soldier 'Bootikins' or Caligula. At the tender age of 24, he became ruler of the Roman Empire. As he processed to Rome for his installation in 37 CE, the populace roared their approval and called him their star, their chick, their babe and their nursling.[47] More than 160,000 animals were consumed over the next three exhausting months of feasting. Is it any wonder he was self-indulgent?

Gaius Caesar Germanicus, or Caligula – his embarrassing nickname by which he is better known – is a byword for excess. Ancient sources (all censorious) tell us his appetite for cruelty and depravity was matched by his appetite for food, sex and luxury. He would shut the granaries and leave his people to starve while he indulged in elaborate feasts. At banquets, he had people tortured as table-side entertainment. During a time when the Roman table groaned under milk-fattened dormice, pike livers, peacock and roasted boar stuffed with live birds, he outdid all other gourmands in his inventive extravagance.[48] He enjoyed elaborate pranks like inviting his dinner guests to sit down to bread and joints of meat – made of gold – while

Unknown artist, *Emperor Gaius, Known as Caligula*,
Roman, 37–41 CE, marble bust.

he swigged back priceless pearls dissolved in vinegar.[49] He yearned
to move the empire's capital city to cosmopolitan, pleasure-loving
Alexandria, where Cleopatra had performed the same feat with pearls
for Mark Antony.

His gluttony for pearls caused his critics some distress. Pliny the
Elder disapproved of his taste for such 'feminine luxuries' as shoes
adorned with pearls. Lollia Paulina, his vulgar wife, dripped with
pearls and emeralds and would show anyone the receipts.[50] Although
he quickly burned through the healthy surplus of funds accumulated
by the emperor Tiberius, his more frugal and reclusive predecessor, it
wasn't his profligacy that led to his downfall. It was his reputed lunacy.

Like Catherine the Great, he was tarnished by the suggestion of
improprieties with horses. Caligula's favourite horse, Incitatus, was
housed in a marble stable with his own furniture and slaves. He had
imperial purple blankets and a pearl-studded collar. Caligula planned
to make him a consul, a position of the highest prestige.[51]

Indulging a horse, though, was a less harmful derangement than
irrational behaviour as a commander-in-chief. Despite Caligula's
army-moniker, he had little military experience and, after mobilizing
his troops for an attack on Britain, floundered at the coast of Gaul.
Ordering his men into formation, he then instructed them to fill their
helmets and folds of their cloaks with seashells before retreating.[52]

His biographer Suetonius includes this episode as evidence of
his 'monstrous' unravelling. But later historians propose this fit of
beachcombing might instead have been a pearl hunt.[53] Resigned to
making the best of a fiasco, Caligula may have told his troops to gather
pearl-bearing mussel and oyster shells. Peering over a gunmetal
North Sea, Caligula and his men might have thought about Britain's
reputation for pearls.[54] The same ocean that lapped the British coast
now lapped at their *caligae* and might deliver the same treasure.

This might have been a symptom of misplaced optimism rather
than madness. Two millennia later, Caligula has undergone a cau-
tious rehabilitation. His paranoia and his mistrust of even his close
supporters turned out to be warranted, given his murder at the hands
of three of his Praetorian Guard after he had reigned for only four

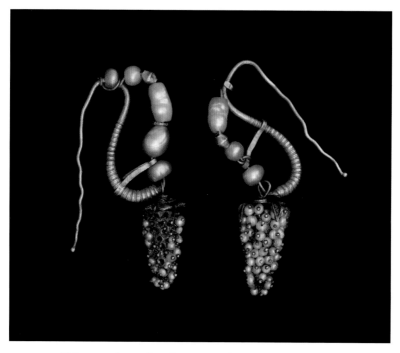

Unknown artist, earrings, Roman, 1st–4th century CE, gold, pearls.

years. During his lifetime his brothers were assassinated and his sisters conspired against him. In a blood-soaked era, tyranny was a survival strategy.[55]

Addiction to opulence and gluttony for pearls were not unusual for the Roman elite. Romans had an endless appetite for pearls. They were studded into furniture, stitched onto clothing, and wrapped around fingers and wrists. They hung in clusters from the ears of Roman women, where they rattled like castanets. They decorated the laces of sandals and even the soles of shoes. They were gathered as the spoils of war and carried in victory processions. On his return from campaigns in the east, the general Pompey had a victory portrait executed in pearls. Pliny disapproved of the egotism but couldn't help admiring the dashing effect of 'hair thrown back from the forehead, delighting the eye'.[56]

In the Roman world, pearls were status symbols, and for an emperor as addicted to spectacle and pageantry as Caligula, they were one more means to brand himself. Gluttony was a political tactic. Lavishing gems on horses was competitive, not manic. The noble Incitatus was a cherished possession, and adorning his horse was the same as adorning his house or his wife. Many societies encrusted horses' bridles, saddles, stirrups and reins with rubies, carnelians, emeralds and pearls.[57] Any horse-loving culture flaunts tack. Caligula might have gorged on pearls but never lost his senses.

Pernille Lauridsen, *Ice Cream*, 14-carat gold, pearl.

ENVY: THE FAIREST OF THEM ALL?

When Mary Stuart returned to Scotland from France as a teenage widow in 1561, she brought with her an enviable stash of jewels. Even stern Scottish Reformer John Knox had to admit, 'Shee brought with her als faire jewells, pretious stones and pearles as wer to be found in Europe.'[58] Pearls, rubies and diamonds decorated her headbands; she had necklaces of unfaceted rubies, table-cut diamonds and large pearls, others with sapphires and pearls; her belts glittered with rubies and diamonds, and dripped pearls.[59]

She also brought with her the famed 'Medici pearls', a thick rope of enormous pearls gifted to Mary in France by her mother-in-law Catherine de' Medici, who had received them herself as a wedding present from her uncle, Pope Clement VII.[60]

Mary returned to her country of birth as Queen of Scots, but also as a monarch who had a strong claim to the throne of England. Her grandmother was King Henry VIII's elder sister, placing Mary next in line to the English throne after Henry VIII's legitimate children. As the Catholic Church considered Henry VIII's marriage to Anne Boleyn unlawful, their daughter Elizabeth's claim was, in Catholic eyes, tenuous.

Three years before Mary's homecoming, she and her cousin Elizabeth Tudor launched a corrosive play of brinkmanship, manipulation and envy that would last the next three decades. In April 1558 Mary married the French Dauphin Francis in Paris. Dressed in white and lavishly bejewelled, the tall, auburn-haired fifteen-year-old was judged one of the most beautiful women in Europe, 'a hundred times more beautiful than a goddess of heaven'.[61] When Mary Tudor died in England later that year, the French monarchy was keen to press Mary Stuart's claim. In France, Mary and Francis were declared the legitimate rulers of England and Ireland. Brazenly, their new coat of arms displayed English lions.[62] Despite their posturing, Elizabeth was proclaimed England's queen in November. But the threat that Mary, Queen of Scots might usurp her throne dogged Elizabeth until she ordered her rival's execution for treason in 1587. Along the way, she acquired Mary's Medici pearls.

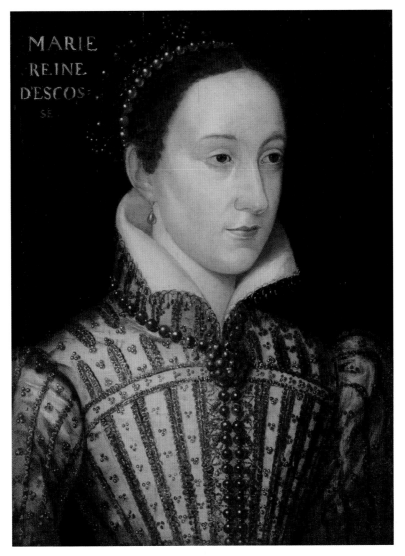

François Clouet (after), *Mary Stuart, Queen of Scots*, 17th century, oil on oak panel.

The story of this intrigue is part of the slow-drip of loss Mary suffered after the pinnacle of her wedding in Notre-Dame Cathedral, when she dazzled in white and gold, blazing with jewels and wearing a crown of diamonds, rubies, sapphires, emeralds and pearls. Less than three years later, Francis died of an infection, and Catherine de' Medici was demanding the return of France's crown jewels. Mary

Unknown artist, *Elizabeth I*, 1550–99, oil on panel.

managed to return to Scotland with the stash of personal property that awed John Knox, but after two disastrous marriages and her abdication, her property began to trickle out of her grasp. Her opponents seized what assets they could and melted her fine silver tableware to mint coins.[63]

She had made her half-brother, the Earl of Moray, custodian of the jewels locked in a coffer in Edinburgh Castle. But as the Regent for Mary's infant son, Moray needed to raise funds fast. Soon Mary's gems were being sold or pawned for cash to float the royal economy. Predictably, some made their way into the hands of noblewomen who likely envied the pearl-studded headbands that once held smooth Mary's famed auburn hair, her pearl and sapphire necklaces, and the pearled belts that enhanced her regal 5'10" stature. Moray's wife, Agnes, acquired a coral belt and a magnificent diamond and ruby jewel in the form of an H, the famous 'Great Harry', which had been a gift from Mary's French father-in-law, Henry II.[64]

Stripped of title, belongings and freedom, by 1568 Mary didn't have much left to envy. Except for her birthright and the support of the pro-Catholic factions who continued to harry Elizabeth's reign. And, of course, her pearls.

Elizabeth, like her father Henry VIII, was a renowned pearl-lover. Portraits show her wearing sinuous ropes that fell below her waist. They were embroidered onto her clothing and wired into hairpieces to create a luminous halo around her head. When the chance came to acquire Mary Stuart's fabled Medici pearls, she grabbed it. The story is related by Jacques Bochetel de La Forêt, the French ambassador to England. In spring 1568 the Earl of Moray sent an agent to London with a selection of Mary's best jewels, including six sumptuous strands of pearls, which he presented to Elizabeth for her inspection. Though the English queen had first refusal, the French ambassador reassured the pearls' original owner, Catherine de' Medici, that she could retrieve them, as Elizabeth was vacillating. He guessed their value at up to 16,000 crowns (over £2.1 million today).[65] But the pearls' magnificence and their royal provenance proved too tempting. Elizabeth knew precisely what she was taking from her rival. Moray was pressed for funds and eager to sell, and Elizabeth secured Mary's pearls for 12,000 crowns (about £1.6 million today).

Both Moray and Elizabeth might have considered the betrayal a small one. Mary had entrusted her half-brother with her jewels, but of larger concern was the governing of a cash-strapped country. Elizabeth claimed she was incensed at the poor treatment of her 'sister' but was always alert for a way to neutralize her threat. The incident of the Medici pearls reflected the toxic sibling rivalry that ended for Moray with an assassin's bullet on an Edinburgh street, for Mary on an execution stage in Fotheringhay Castle, and for Elizabeth in London, crying crocodile tears for the cousin whose annihilation she helped engineer.[66]

Some of Mary's jewels, particularly the ones that had been pawned, were eventually recovered. Even Agnes, Countess of Moray, had to give up the 'Great Harry'. But the Medici pearls remained with Elizabeth.

The motivations behind Mary's murder were complex, but envy played a role – even if it wasn't quite the driver that popular culture, such as Josie Rourke's 2018 film *Mary Queen of Scots*, would have us believe. (During a fictional meeting in Rourke's film, Elizabeth admits to Mary, 'I was jealous. Your beauty, your bravery. Your motherhood.') Elizabeth seems to have hesitated to deprive Mary of her pearls, just as she hesitated to deprive her of freedom then of her life.

François Clouet, *Catherine de' Medici*, c. 1555, watercolour on vellum.

For her execution, Mary wore a dress of black satin trimmed with pearls.[67]

> Pearl queen –
> she imagines the hall is filled with pearls
> and the pearls are snowdrops.
> Into the sheer purity of it she will fall.

> She is going into the arms
> of her mother.[68]

GREED: BAJA MIRAGE

John Steinbeck's 1947 novella *The Pearl* is a parable about the danger of sacrificing integrity for riches. Set in the coastal town of La Paz in the Mexican state of Baja California Sur, the story traces the tragedy of a young indigenous family: Kino, his wife, Juana, and their infant, Coyotito. After the baby is stung by a scorpion and the town's doctor refuses to treat him, his desperate parents paddle their canoe into the bay to try their luck at pearl fishing. Kino miraculously retrieves a large pearl as perfect as the Moon. Before long, the news of his discovery seeps through the small town, and the family begins attracting unwelcome attention. Wary of being cheated by the town's pearl-buying agent, Kino resolves to take his pearl to the capital city to secure a higher price. But before he can depart, he is attacked, and in the struggle, kills his assailant. The family flees on foot with their baby into the mountains, pursued by trackers on horseback. In a desperate climax, Coyotito is shot dead by one of the trackers. Having lost everything they own – their home, means of livelihood, and their firstborn son – Kino and Juana return to La Paz and hurl the pearl back into the ocean.

But if this is a story about greed, it wasn't Kino's. Steinbeck was well-acquainted with Mexico's landscape, history and the harsh circumstances endured by its indigenous peoples. He visited the country frequently and lived for several months in Mexico City. In the spring of 1940, he and the biologist Edward Ricketts chartered a sardine fishing boat, the *Western Flyer*, and spent six weeks collecting marine specimens along the Gulf of California. Also known as the Sea of Cortez, this inlet of the Pacific separates the Mexican mainland from the slender finger of mountainous land that points 1,200 kilometres (750 mi.) south from the United States/Mexico border. They documented their trip in the book *Sea of Cortez*.[69]

The trip was ostensibly a chance to study the region's marine life, though as Steinbeck tells it, it became an adventure at sea, an exploration of sterile desert islands and odiferous mangrove swamps, an opportunity to examine not just starfish and anemones but also the human soul. Anchoring the *Western Flyer* off-shore, they would guide their 'irritable' little outboard motor, the *Sea-Cow*, to coral reefs, rocky shorelines and sandy coves. They slogged across great flats of eelgrass in rubber boots, poked around rockpools 'ferocious with life', peered at millions of sulphury sea cucumbers curdling in the heat. They saw great thrashing schools of tuna, bloated pufferfish, hammerhead sharks, Sally Lightfoot crabs scuttling on their tip-toes, huge stalk-eyed conch, fiddler crabs, starfish, limpets and sponges.

But they also saw the precarious existence of the region's indigenous inhabitants, wresting a living from the rocky coastline, which butted hard onto arid desert. Steinbeck had read historical accounts of the violence of the region's European settlement and knew that life, in this gulf of seductive turquoise bays, hot yellow sand and treacherous currents, was too often blighted. The indigenous inhabitants are a shattered and distrustful background presence in Steinbeck and Ricketts's *Sea of Cortez*. At Pulmo Reef on the peninsula's southern tip, they were approached by a small group of indigenous people, *serapes* so thin they were see-through. One offered them a matchbox with 'a few misshapen little pearls like small pale cancers'. The travellers accepted them in exchange for a carton of cigarettes.[70]

The Spanish conquistadors in Baja California had been motivated more by avarice for the coastline's fabled pearls and gold than for the salvation of indigenous souls. The region's inhabitants had harvested pearl oysters for over a millennium, and middens of bleached

Antonio García Cubas, map of Baja California from *Atlas Mexicano* (1886).

shells lay scattered throughout the shoreline and islands.[71] Rumours of the peninsula's riches lured early expeditions to this region of bluffs and rock, cactus-studded desert and rust-red mountains. Having conquered Mexico for Spain, Hernán Cortés turned his attention to remote Baja after receiving reports about Pericúes, a people who braided their long hair with 'beautiful pearls'.[72] The settlement that Cortés tried to establish at La Paz in 1535 was a failure, but that didn't dissuade subsequent adventurers, both Crown-sponsored and private. Later that century, entrepreneurs were competing for licences to develop pearl fisheries along the gulf.

In 1596 the Spanish soldier Sebastián Vizcaíno, who had been granted a major concession, sailed to the peninsula with three ships and their crews, plus horses, arms, soldiers and Franciscan friars.[73] They found the Baja coast sandy, hot and rough, the people unclothed and uncivilized. Facing a shortage of drinking water and food, the soldiers threatened to mutiny. Catastrophe piled on catastrophe – one ship was wrecked, the explorers lost their way among strange islands in furious storms, they were attacked by indigenous warriors, and their camp burned to the ground. Conceding defeat, they limped back to the Mexican mainland. They had lasted all of three months on Baja.

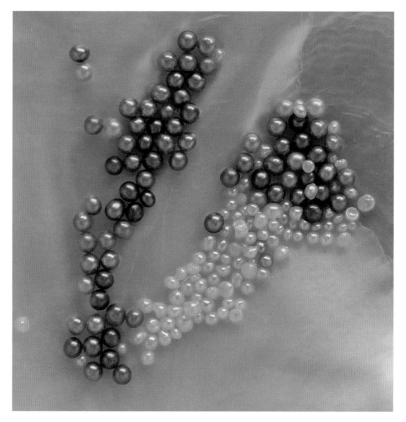

Natural Baja pearls, 1.6–2.2 mm diameter, next to white and cream freshwater cultured pearls.

The only pearls they encountered were a few small ones brought to them by locals more inclined to share the region's small yellow plums.

Despite the heat that shimmered off the ocean and the mirages that made islands out of headlands, despite the cactus-, thorn- and boulder-strewn desert, and the wrinkled hills rising to bare spines of mountains, Baja California continued to draw explorers, pirates and pearl-hunters. *Pinctada mazatlanica* oysters yielded enough large, silver or lavender-hued pearls to make the geographic hazards worthwhile. During the seventeenth century, the gulf became a major pearling centre; Baja pearls adorned the bodies of Spanish queens and the crowns of the Virgins in Spanish cathedrals. The Spanish crown claimed its unpopular *quinto* (20 per cent tax on precious commodities) or failed to claim it from clandestine traders. But if the pearl fisheries could be bountiful, they could also turn up bust. Not a few hopefuls went from cove to cove searching for pearls only to return to the Mexican mainland, 'lost and without glory'.[74] It was as though the Sea of Cortez induced a pearlescent trance. Steinbeck and Ricketts still felt its pull: 'The very air here is miraculous, and outlines of reality change with the moment. The sky sucks up the land and disgorges it. A dream hangs over the whole region, a brooding kind of hallucination.'[75]

The Jesuits arrived in Baja California in 1697, and over the next seventy years of colonization, established a string of missions. The most relevant figure for Steinbeck was Eusebio Kino, after whom he named *The Pearl*'s central character. This mission-founder extraordinaire also charted the first accurate maps of Baja California, published in 1705, documenting that it was a long peninsula, not an island.[76]

A lucrative pearl industry developed in sheltered La Paz Bay, but rapacious fishing severely stressed the oyster beds, which cycled through periods of exhaustion and recovery. Greed undermined the industry's sustainability, as well as the health and well-being of indigenous divers. During the nineteenth century, licensed pearling fleets made up of steamboats and canoes plied the coastline, employing skilled indigenous free-divers. Entrepreneurs could realize great profits, but the divers themselves often remained in penury as they

Illustration from Albert Gilliam, *Travels Over the Table Lands
and Cordilleras of Mexico, during the Years 1843 and 44* (1846).

were paid upfront in food and then struggled to repay their debts. Although they could sell the pearls they found, their employers had the first option to buy, and the terms were not generous.[77]

By the 1930s the oyster beds were so degraded that an environmental trigger in 1939 (possibly a shift in ocean salinity, temperature or pH value) culled the remaining molluscs. Divers found them dead on the ocean floor with their valves open, and soon their empty shells hosted miniature reefs of anemone. In 1940, the year Steinbeck and Ricketts visited La Paz, the government finally declared a ban on pearl fishing.

Steinbeck set his novella in La Paz's fairly recent and troubled past. In 1900, despite indigenous peoples making up 40 per cent of Mexico's population, the property-owning class was mostly white and of Spanish descent.[78] The novella exposes the vast inequities between the two groups. Steinbeck loosely based his story on a folktale he had heard during the trip. This folktale, too, was set in La Paz, the port that supplied pearls for the robes of kings and bishops, but in which 'the terrors of greed were let loose'.[79] In the folktale, a young indigenous diver finds an unbelievable pearl and dreams of escaping his impoverished life:

In his one pearl he had the ability to be drunk as long as he wished, to marry any one of a number of girls, and to make many more a little happy too. In his great pearl lay salvation, for he could in advance purchase masses sufficient to pop him out of Purgatory like a squeezed watermelon seed. In addition he could shift a number of dead relatives a little nearer to Paradise.[80]

As in Steinbeck's novella, when the folktale's diver tries to sell this miraculous stone, greedy brokers deny him a fair price. He is mugged and tortured when he tries to flee inland. In despair, he skulks back to the beach 'like a hunted fox' and hurls the pearl back into the sea, resigned to a future of hunger and hardship.

Pearl fishing, La Paz, Baja California.

The corrupting influence of a pearl was at the heart of the book Steinbeck based on this folktale, but Kino's motivations are quite different from those of the folktale's young diver. Kino needs money not to pay for high living but to cure his infant son. He doesn't find the gem by accident; he finds it after Juana prays that they might find a pearl on the bed 'that had raised the King of Spain to be a great power in Europe in past years, had helped to pay for his wars, and had decorated the churches for his soul's sake'.[81] When Kino finds one as large as a seagull's egg and fantasizes about what it will buy, he imagines kneeling at the high altar of a church and receiving with Juana the sacrament of marriage. Juana imagines her baby baptized.

Like the boy in the folktale, Kino was deceived by La Paz's cartel of pearl brokers, just as indigenous divers had been cheated for so long by licensed pearlers. Kino's neighbours finally deduce the agents' deceit, incredulously grasping, 'If that is so, then all of us have been cheated all of our lives.'[82] And rather than dreaming of jail-breaking his relatives from Purgatory, Kino dreams of sending his son to school so that he can learn the skills he needs to break free of ignorance. The pearl is salvation not from Purgatory but from the greedy doctor in Parisian red silk, from the condescending priest, from the pearl-buying agent with his clean nails and dirty soul, and from the centuries-long shadow of the king of Spain.

It isn't greed that destroys Kino, but aspiration. The priest tells him that rising above one's station is a crime against religion. While the news of Kino's pearl stirs up 'something infinitely black and evil in the town', in Kino it kindles an incandescence.

SLOTH: SHORT CUTS TO DESTRUCTION

Avarice decimated the La Paz oyster beds long before Steinbeck and Ricketts arrived. But sloth contributed. The bleak trajectory of colonial pearling in Baja California traces the European enterprise to expend as little personal effort as possible to extract a valuable natural resource. Gonzalo de Francia, ship's officer for the Spanish soldier Sebastián Vizcaíno, suggested their abortive 1596 expedition failed because Vizcaíno didn't attend to labour-saving inventions:

> the conquest of this country is very important, because
> it has signs of great prosperity, the principal one being
> the many pearl oysters in which there can be much wealth
> once rich beds are discovered. Sebastián Vizcaíno made no
> effort to hunt for them as he carried no divers nor apparatus
> for diving.[83]

There were riches within grasp, but there had to be an easier way to secure them than relying on the region's indigenous people. Before the arrival of the Spaniards in the sixteenth century, Aboriginal peoples numbered around 60,000, and several distinct groups occupied territories throughout the peninsula and its islands.[84] Some were semi-nomadic, travelling between inland oases and the coast, gathering foods according to season and stewarding local ecosystems. Analysis of ancient shell middens reveals that they harvested the older oysters, safeguarding the sustainability of oyster colonies.[85]

The Spanish quickly found that these Aboriginal peoples were less tractable than the divers who worked for their fisheries on the Pearl Coast along eastern Venezuela and its islands. The Spanish explorer Pedro Porter, writing of his Pacific voyage of the 1640s, expressed a long-standing and common prejudice against Baja California's inhabitants who, though skilled divers, 'could not be forced to dive because they never submit'.[86] Faced with a shortage of exploitable labour, Spanish enterprises began importing enslaved black divers to Baja California. By the seventeenth century, these divers were working beside indigenous divers brought over from the Mexican mainland.[87]

Labour-saving devices seemed another logical solution. Francisco de Ortega, who explored Baja California during the 1630s, brought a rudimentary wooden diving bell. Weighted with lead, it could descend with two people, though the mechanism for supplying it with fresh air was not recorded. He also brought three iron rakes, which weighed up to 115 kilograms (250 lb), to dredge oysters from their beds. This second invention, though environmentally destructive, might have been more successful. Based on the recorded amount of pearls relinquished in tax, he recovered at least 5 kilograms (11 lb) of pearls over fifty days' fishing in 1632.[88]

During the missionary era (1697–1740), Jesuit friars prohibited the harvesting of pearl oysters and the pressing into service of the peninsula's indigenous peoples. The health of the oyster beds briefly recovered, but economic opportunities were to stress them to exhaustion. After 1821, a newly independent Mexico began attracting foreign investment, with Britain leading the way. Eager to profit from the country's economic development but wary of sinking capital into risky propositions, companies sent agents on reconnaissance tours to assess the profitability of proposed ventures. Lieutenant Robert Hardy, who was dispatched to Mexico during the 1820s by the General Pearl and Coral Fishery Association of London, wrote a revealing account of the economic calculations behind the exploitation of the gulf's pearl oysters. He observed that pearling operations were currently stripping the beds of all oysters, regardless of size or age, and that the stock was dangerously depleted. Although oysters existed at depths unreachable by free-divers, he judged that diving bells would be useless because the fissures and chasms on the ocean floor would leave the bells' occupants 'suspended, half-way between hopes and realization'. The unhealthy state of the oyster beds and the expense of fitting out vessels led Hardy to conclude that his employer should have nothing to do with 'so wild an enterprise as that of the Mexican Pearl Fishery'.[89]

Hardy went to commendable lengths to provide the Association with an accurate assessment of risk. To understand the exertions demanded of the pearl diver, he made several dives himself to depths of 6 metres (20 ft), 'at which depth the pressure of the water upon the ears is so great, that I can only compare it to a sharp-pointed iron instrument being violently forced into that organ'.[90] Assured that once his eardrums burst he would have an easier time, he persisted, tormented as much by his mind as by physical pain:

> Every fathom fills the imagination with some new idea
> of the dangerous folly of penetrating further into the silent
> dominions of reckless monsters, where the skulls of the dead
> make perpetual grimaces, and the yawning jaws of sharks
> and *tintereros* [sharks], or the death-embrace of the manta,
> lie in wait for us. These impressions were augmented by the
> impossibility of the vision penetrating the twilight by which
> I was surrounded, together with the excruciating pain I felt
> in my ears and eyes.[91]

At a depth of 10 metres (33 ft), struggling through jelly-dense water, he felt an explosion in his head. He surfaced with blood streaming from his ears, eyes and mouth.[92]

It was little wonder that companies tried to find machines to re-place free-divers. One New York-based company touted a Submarine Explorer that would do the work of fifty men.[93] No serious investors expressed interest, but the introduction of the diving suit in 1874 cemented the capitalistic nature of pearl diving. Well-funded private companies were granted concessions of vast areas of the Gulf, which was carved into privatized marine spaces. The divers themselves were poorly paid, and their working conditions harsh. One of the most predatory of these companies was the British-owned Mangara Exploration Limited Company, which by the early twentieth century had rights to most of Mexico's pearls. Subjecting its divers to debt, unsanitary housing and hazardous working conditions, the company was also a scourge on the environment, using devastating methods such as dynamite, dredges and trawlers.[94] While this recklessness stripped the ocean floor, one positive outcome was a new focus on aquaculture through the pioneering work of La Paz native Gaston Vives. In the early 1900s, his breeding station on Isla Espíritu Santo in the Bay of La Paz was the first to mass culture pearl oysters.[95]

As Robert Hardy discovered for himself, pearl-diving was a physically and mentally arduous activity. For hundreds of years, Baja California's indigenous groups had sustainably farmed shellfish in the Gulf of California. The decimation of their people mirrors the

Fritzia Irízar, *Untitled (Mother-of-Pearl Graft)*, Mexico, 2015–18,
artist's grafting of symbols of currency into indigenous Baja oysters.

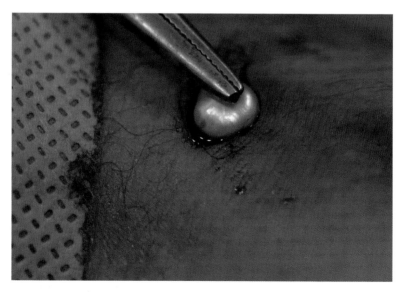

Fritzia Irízar, *Untitled (Human Pearl Graft)*, Mexico, 2019,
video depicting a pearl being surgically implanted into
human arm as a critique of extractive capitalism.

depletion of the Gulf's pearl oysters. From pre-contact figures of around 60,000, the Aboriginal population collapsed, until by 1835 only about 1,000 inhabitants remained.[96] Imported diseases, some introduced by pearlers, had taken a catastrophic toll.[97] Ensuring the sustainability of pearl oyster populations required the labour and knowledge of peoples who no longer existed. But destroying a reef took all the effort of detonating a charge of dynamite.

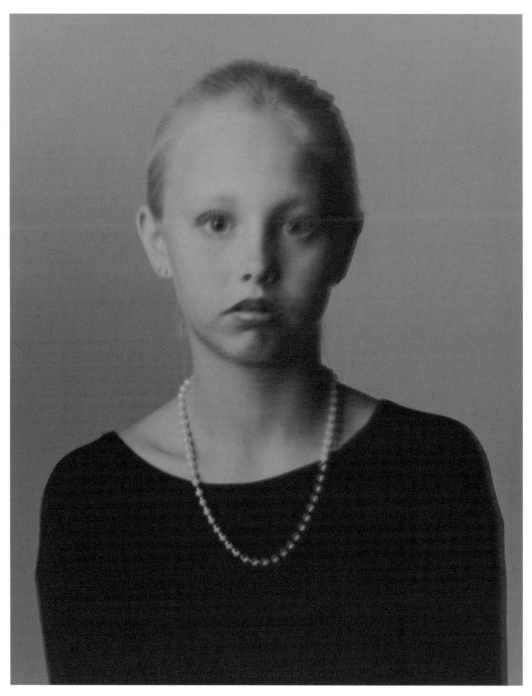

Robert Neal Stivers, *Young Woman #1*, 2001, gelatin silver print.

❦

FIVE

And Seven Virtues

Pearls carry the weight of our values and aspirations, our desires and dependencies. Like so many objects we decree precious, they've turned us to murder, mayhem and madness, tugging us down through each circle of Hell. But like Dante's muse Beatrice, they also point us heavenwards. In *Paradiso*, the first star Dante and his guide reach is the Moon – 'the everlasting pearl' – that receives them 'just as water will accept a ray of light and yet remain intact'.[1] Moonlight inflames the poet's longing for union with God, just as a pearl's clean sheen might reflect our better selves.

CHASTITY: THE SACRED PEARL

The medieval English masterpiece *Pearl* displays the gem's meanings at their most elastic. The 101-stanza poem relates a father's heartbreak over the death of his young daughter. Returning 'hollow with loss and harrowed by pain' to the garden where he last saw his child, he collapses in grief and dreams that she is restored to him as a radiant young woman.[2] Persuaded by his longing that she is just within his grasp, he lunges towards her but awakens to find himself once more bereft.

The pearl is the perfect metaphor for the girl. Under two years old when she died, she is 'so small, her skin so very smooth', spotless and priceless.[3] Her apparition, crowned with pearls and dressed in luminous white linen stitched with the gems, is pearl adorned, pearl-like and a transfigured heavenly pearl of great price. 'Not stained or shamed by corrupting sin', she has become a bride of Christ, 'a pure and virgin queen'.[4]

Pearl was probably written in the 1390s when associations between pearls and virginity were commonly accepted. Oysters themselves are gloriously fecund, rhythmical hermaphrodites that switch sex in response to fluctuating ratios between male and female.[5] They spawn en masse, ejecting clouds of sperm and eggs that can turn water milky, a frenzy that seems the opposite of chastity. Yet the idea that an oyster decorously bobbed to the water's surface, to be impregnated by dew or a moonbeam to make a pearl, clung on for centuries.[6] Venus, herself a pearl, was born pure from Cypriot seafoam. Chaste Artemis was goddess of the pearl-white Moon.

The Virgin Mary's pearl-like nature stemmed from her virginity and her immaculate conception – being born free of sin. That piece of theology had been debated since the fourth century, but as the dogma became more entrenched, artists began to deck out Mary in pearls. The stones' connection with virginity reached a visual sublime in the spectacle of Queen Elizabeth I of England and Ireland. Following the Reformation, which suppressed images of the Mother of God, Elizabeth cannily appropriated the Marian cult for her own agenda. An early adopter of fashion as propaganda, her dress signalled her chastity and worthiness to assume the male office of kingship.[7] Like the young daughter in *Pearl*, her virtue is fashioned through dress. In portraits, she dazzles in swags of pearls; they halo her auburn curls. Squadrons of seamstresses spent their days transferring jewels between her 3,000 gowns, painstakingly stitching drilled pearls to silks and satins.[8] Stiff with ornament, this clothing denied rather than enhanced her femininity. It flattened her breasts, hid her hips, and immobilized her throat in a starched ruff. The queen's body was simultaneously offered to her subjects and erased under armour. 'Such heat in ice, such fire in frost,' wrote Sir Walter Raleigh in his long love poem to the queen, *The Ocean to Cynthia*.[9] Written while

Unknown artist, pendant brooch with cameo of enthroned Virgin and Child, Byzantine,
late 11th to 12th century (cameo), 12th to 14th century (mount), chalcedony cameo;
gold mount with pearls, emeralds, garnets, sapphires and a sardonyx intaglio.

Raleigh was in disgrace, the poet pines for the favour of a monarch as
untouchable as the Moon – of which Cynthia (Artemis) was goddess.

In Nicolas Hilliard's *Phoenix Portrait* of circa 1575, Elizabeth's
lush black costume is a pincushion studded with white pearls. She
is a pale Virgin Queen married to the state. Between her breasts
is a phoenix, long understood as a symbol of Christ's resurrection.
Once again, costume, amplified by Elizabethan oratory and poetry,
transports the woman into the realm of the spiritual. She is 'a phoenix
rare', sacrificing herself for her nation and, despite birthing no heirs,
ensuring the redemption and longevity of the state.[10] She is not just
Marian but Christ-like.

The poem *Pearl* makes the same leap between the Virgin and
Christ but far eclipses the complexity of the *Phoenix Portrait*. The

Nicholas Hilliard, *Queen Elizabeth I* (known as the *Phoenix Portrait*), *c.* 1575, oil on panel.

pearl is the blameless human child and the spirit transformed into the Bride of Christ. But, just like Elizabeth I is both Virgin and Saviour, the child's soul, joining the lamb-white, pearl-white body of Christ, becomes Christ.[11] The early Christian Church conceived of Jesus as a pearl in the shell of Mary's womb. This also happened to be an era of lavish pearl-wearing, leading the theologian Clement of Alexandria (*c.* 150–*c.* 215) to sputter against women adorned 'with a sacred stone, the Logos of God . . . the serene and pure Jesus'.[12]

The *Pearl* poet goes further. Invoking the parable of the pearl, in which a merchant trades everything he values for a precious pearl, the gem mineralizes into the Kingdom of God. Early in the poem, the father compares himself to a jeweller searching for his lost treasure. His daughter chides him – 'quit your carping' – God is no thief. By the poem's conclusion, the father's brief vision of the pearl-gated New Jerusalem has been snatched away, but his consolation is that the Kingdom of God lies on earth, whenever people live 'both as His lowly servants and beautiful pearls, pleasing to Him'.[13]

A haunting meditation on loss and theological discourse on death, *Pearl* is also a technical marvel. Beginning and ending in a garden, the poem's 101 stanzas are a faultless circle. They echo the 'endless sphere' of 'the luminous empire of heaven'.[14] The stanzas are linked by words repeating from closing to opening lines, a linguistic conchiolin cementing the shimmering layers of nacre. Each line glitters with alliteration ('from every angle you are angel-like').[15] The poet is a jeweller, tooling an intricate Fabergé egg of nested allusions. And in becoming a jeweller, the poet imitates God, who creates the Word incarnate and crafts the kingdom that lies beyond grief.

HUMILITY: THE WORM AT THE HEART OF THE PEARL

The tiny shoulder bone of a Scottish queen who died more than nine hundred years ago fits in the hand like a bar of soap. People were, of course, much smaller in the Middle Ages, but the owner of this diminutive scapula was likely malnourished.[16] It belonged to St Margaret (*c.* 1045–1093), the English princess who became queen of Scotland on her marriage to Malcolm III, slayer of King Macbeth. It's the only piece of her left, though her fingers and skull and shards of her thigh bone rattled around Britain and Europe for centuries. At one time, King Philip II of Spain owned a nerve from her leg, a relic valuable enough to list in a royal inventory of the Escorial Palace.[17]

The malnutrition is a puzzle because Scotland in the Middle Ages teemed with wildlife. A queen's table would have buckled with venison stews, roasted boar, fresh oysters, mussels from cold rivers, grilled salmon and brown trout, berry tarts, custards, baked pears and spiced gingerbread. All washed down with honey mead. The only reason she had to go hungry was that she chose to.

A decade after her death, her spiritual advisor Turgot, prior of Durham Cathedral, wrote an unctuous biography for her daughter, Edith-Matilda. Playing on Margaret's name, derived from the Latin for 'pearl', he enthused, 'the fairness indicated by her name was surpassed by the exceeding beauty of her soul . . . and in the sight of God she was esteemed a goodly pearl by reason of her faith and good works'.[18] Through the purity of her devotion, Margaret embodied the perfection of the pearl. Her life was 'a pattern of the virtues', but Turgot was at pains to stress the greatest of these was her humility.[19]

As a queen, Margaret had, like Cleopatra, every reason for humility's opposite, pride. She could claim an impeccable lineage; she was related through her father to Edward the Confessor. As 'the noblest gem of a royal race', she brought glamour to a dowdy court. She welcomed traders from overseas who seasoned the Scottish diet with exotic spices and herbs. A cosmopolitan fashion maven, she persuaded the Scots nobility to discard their dun-coloured woollen clothes for fabrics saturated in bright dyes. Wide-eyed, Turgot remarked, 'they might have been supposed to be a new race'. She built churches filled with gold, silver and gems and patronized the arts. Improbably, Scotland acquired the hummingbird sheen of conspicuous consumption.[20]

To honour God and enhance the prestige of her new homeland, Margaret insisted on splendour; she dedicated her personal life, though, to pious asceticism. Turgot, writing within the living memory of her family and court, claimed that she repressed 'all swellings of pride' and refused to let flattery 'fatten her head'. The opposite of epicurean Cleopatra, she fasted from childhood, eating 'only to sustain life and not to please her palate'.[21] Like someone with an eating disorder, she fixated on food. As Turgot observed, each scant mouthful stimulated her appetite instead of satisfying her hunger. Half-starved, she sought ways to sharpen the agony. Her ladies brought her small orphans (Turgot fails to mention if they were bathed and re-dressed), and she sat them on her bony lap and fed them with her own spoon. Maybe she imagined the taste of each mouthful, the delicious beads of oil slipping down her own throat. Her asceticism turned her wan. Inevitably, she destroyed her health and died before she reached fifty. Ironically, she has been feted through history for introducing cutlery to the Scottish table.

Turgot's Margaret emerges as a woman so pious, self-abnegating and devoted to family and nation that it wasn't long before people clamoured for her canonization. For years after her death in 1093, pearly light shimmered from her tomb in the church she had founded in Dunfermline, and she appeared in visions to the resident monks.[22] The papacy, ever-vigilant for frauds and trick miracles, at last relented, and Pope Innocent IV made her a saint in 1249. When her corpse was exhumed to place in a new tomb in June 1250, observers were relieved to breathe air perfumed with 'the fragrance of spices and the scents of flowers in full bloom'.[23]

Her miracles extended to women in childbirth (she had eight children who lived to adulthood), and her chemise was a comfort

Site of the shrine of St Margaret, Dunfermline Abbey, Fife, Scotland.

to pregnant Scottish queens.[24] Nearly five centuries after Margaret's death, Mary, Queen of Scots ordered her head brought to Edinburgh, where it protected her as she delivered her son, James VI.[25] The smooth skull with its flowing auburn hair saved Mary through this ordeal and for several years afterwards. But it couldn't save her from losing her own head on a blood-soaked block at Fotheringhay Castle.

While Mary died a traitor, Margaret expired a paragon of virtue, clutching a pearl-studded cross. 'This pearl, I say, was taken from the dunghill of this world and now shines in her place among the jewels of the Eternal King,' exclaimed Turgot.[26] His reference to ordure is telling, as it inadvertently shone a light on the very thing she abhorred – our animal selves. For Margaret, the greater the appetite she resisted, the brighter her glow. She repressed her own faults and those of her children, exhorting their tutor to scold and beat them when they became too high-spirited. She reformed the church, pruned out pagan practice, extended Lent and cleansed the temple. It was as though she understood what science would take centuries to uncover: that a worm might coil in the heart of the pearl.

PATIENCE: ENDLESS TEARS

If wrath culls life, patience, its opposite, makes us immortal. For centuries, reliquaries enshrined locks of hair and shards of bones to embody saints' souls and memories of their lives and teachings. Mourning jewellery, which became increasingly popular in England in the Early Modern period, was the secular version. Preserving the beloveds' names, the dates of their lives and sometimes their hair, these comforting objects denied death's finality and eased the wait for reunion.

Initially, though, creating lockets and rings to commemorate the dead was as much about social status as sentiment. Samuel Pepys's will provided for 123 mourning rings of three different qualities to be gifted at his funeral in 1703.[27] While collecting these rings became a fad, high infant mortality rates made such mementoes a sad necessity for those who could afford them. In 1638, after the death of his young daughter, Anna Maria, the English baronet Sir Ralph Verney wrote to his brother promising, 'you shall herewithal receive a ringe filled with my deare gerle's hair; she was fond of you . . . therefore I now send this to keep for her sake'.[28] Around 1800, mourning jewellery in Europe and America began featuring pearls, often as a border around hair mementoes of the deceased. These pieces employed a shared language of loss, with pearls taking their place in the syntax. Like hair, pearls, under the right circumstances, last forever, thwarting the sting of decay. They confirm the existence of eternity.

Pearls also had millennia of accumulated symbolism connecting them with sadness. Their harvest requires the sacrifice of an animal, so they hold death at their core, each carrying the trace of its maker. Kōkichi Mikimoto's elegiac mounds of empty oyster shells are a glimpse into the scale of the obliteration. Before a biological understanding of pearl formation, world myths intuited the melancholy by claiming that the tears of angels, goddesses, nymphs, sharks or mermaids beget pearls.[29] The equation between pearls and tears allowed Shakespeare's Richard III to goad the mother of the children whose murder he had orchestrated: 'The liquid drops of tears that you have shed/ Shall come again, transformed to orient pearl' (IV.4). The trope was so firmly cemented that Shakespeare could often omit 'tears' entirely, writing, 'those heaven-moving pearls from his poor eyes' (*King John*, II.1), or 'Those round clear pearls of his, that move thy pity' (*The Rape of Lucrece*, l. 1553) or 'What guests were in her eyes, which parted thence/ As pearls from diamonds dropped' (*King Lear*, IV.3).

If, metaphorically, salty tears fused with marine pearls, it was a small step to flip the myth of tears making pearls into a belief that pearls forecast or even caused grief. In John Webster's violent tragedy *The Duchess of Malfi*, the duchess relates a disturbing dream to her husband, Antonio:

> DUCHESS: Methought I wore my coronet of state,
> And on a sudden all the diamonds
> Were chang'd to pearls.

ANTONIO: My interpretation
Is, you'll weep shortly; for to me the pearls
Do signify your tears. (III.5)

As the omen unfolds, she is hounded, imprisoned, tortured and finally strangled.

The duchess's transgression was remarrying against her family's wishes. Pearls have played a curious role in the marriage ceremony. Virginal white, they are an appropriate love token, but simultaneously, they retain their associations with doom. On her wedding day to Emperor Napoleon III in 1853, Eugénie de Montijo was implored not to don a long pearl necklace.[30] Despite its glittering highs, Empress Eugénie's life was a catalogue of heartbreak. Deposed in 1870 (a fate seemingly presaged by her obsession with Marie Antoinette), she was exiled to England and lost her husband and only child within a decade. The French Crown Jewels, including lavish quantities of her favourite pearls, were auctioned by the state in 1887.[31] Carl Fabergé bought the magnificent pear-shaped 'La Régente' and sold it to a Russian princess.[32] In exile, Eugénie's beloved personal pieces trickled away to raise funds for her maintenance. The American industrialist George Crocker snapped up the magnificent grey pearl and diamond earrings that she wore for an 1854 portrait by Franz Xaver Winterhalter.[33]

The next century, the Spanish dramatist Federico García Lorca captured pearls' dark spell in his claustrophobic play, *The House of Bernarda Alba* (1936). When the aptly named Angustias (anguish) shows her engagement ring to a family friend, the latter responds, 'In my day pearls signified tears.'[34] To make matters worse, Angustias's sister has just spilt the salt. Before the act's end, resentment and sexual jealousy cascade into violence and suicide.

Pearls' equation with sadness transformed them into the perfect mourning jewel. Secreted in museum collections around the world are treasures that bear enduring witness to loss and remembrance. One ring now in the Walters Art Museum, Baltimore, records a multiple blow. Its white enamel band enshrines the innocence of

Franz Xaver Winterhalter, *The Empress Eugénie (Eugénie de Montijo)*, 1854, oil on canvas.

Unknown artist, memorial ring, British, 1804–5, gold, enamel, pearls, hair.

two unmarried siblings. The gold inscriptions document the double catastrophe, separated by less than eight months: C. M. BURNLEY. DIED. 3 MAR. 1804. AGED 19. A. E. BURNLEY. DIED. 8 JULY 1803. AGED 20. A pearl-mounted bezel frames their hair, the younger sibling's just one shade lighter than the elder's. Such are the shades of grief.

As mourning, especially for women, became more codified during the nineteenth century, pearls took their place in a rigid system of decorum. A woman might spend a year dressed in monotone black in full mourning for a close family member. After that, she could progress into half-mourning, when she could wear pearls and a mauve or brown dress.[35] The gems' lustrous white surfaces contrasted with the beloved's hair smoothly braided and set into lockets and rings. Sometimes, tiny seed pearls picked out initials. During the sentimental Victorian era, hair work surged in popularity, practised by women as a hobby at home, as a cottage industry and eventually on an industrial scale.

Unknown artist, *Queen Victoria*, 1898, photograph in red plush frame.

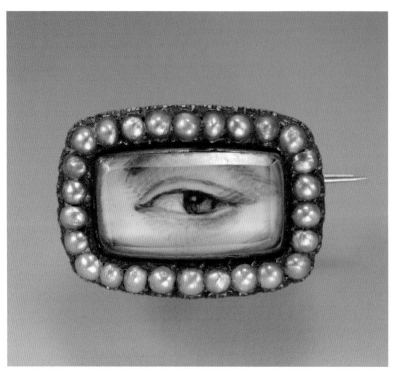

Unknown artist, 'eye' brooch, British, early 19th century,
portrait miniature, pearls, rock crystal, gold.

Britain's most illustrious mourner, Queen Victoria, pined over
the loss of Prince Albert for four decades, remaining in austere black
from 1861 to the end of her life. She had jewellery made with his hair
and held seances to try to reach him. 'I feel now to be so acquainted
with death,' she explained to their daughter, 'and to be much nearer
that unseen world.'[36] Her black silks signalled grief, but her pearls
imparted patience. They circled her throat and wrists, each pristine
orb capturing in their roundness the promise of unbroken love, the
last breath in this world exhaling into the first breath in the next.

TEMPERANCE: PEARLS FOR A YOUNG REPUBLIC

In 1802/3, the poet Sarah Wentworth Apthorp Morton sat in
Philadelphia for the acclaimed portraitist Stuart Gilbert. His finished

painting depicts the writer with the tools of her craft – inkwell, papers and quill pen – watched over by a sober bust of her friend, George Washington. She wears a dark dress embellished with a modest ruffle of lace and three demure strands of white pearls around her neck. Around her left hand, she is fastening a gold bracelet with pearls. The restrained elegance of her jewels perfectly fits her persona as a woman of letters engaged with the pressing concerns of the new Federal Republic.

Sarah Apthorp was descended from a family of prosperous New England merchants. A member of the educated elite, in 1781 she married the ambitious patriot Perez Morton, Harvard-educated scion of an old Massachusetts family. The couple assumed their assigned roles in Boston's social circles, and Perez Morton continued his political rise. He was elected to the Massachusetts House of Representatives in 1794 and eventually to the position of Attorney General of the Commonwealth, a role in which he served for twenty years.[37] His wife, meanwhile, carved her literary career. Grappling in print with the major political and cultural issues of the era – the abolition of slavery, women's rights, the French Revolution, the situation of America's indigenous peoples – she became not just a bluestocking salonnière, but a public intellectual.

Her first major poem, 'Ouâbi, Or the Virtues of Nature: An Indian Tale', was published in 1790. Although it romanticizes the sacrifice and loss endured by indigenous peoples, the work was much admired by Sarah Morton's contemporaries and even adapted for the stage. Its sympathy for Native American experience was also an overt critique of the dissipation and 'European vices' the new Republic was leaving behind.[38] Opening in a 'black forest and uncultur'd vale', its setting and values are the opposite of gluttonous indulgence:

> 'Tis not the court, in dazzling splendor gay,
> Where soft luxuriance spreads her silken arms,
> Where gairish fancy leads the soul astray,
> And languid nature mourns her slighted charms.[39]

Shortly afterwards, when the British Parliament failed to abolish slavery, Morton responded in print. Her poem 'The African Chief' (1792), which describes the death of a black slave during an uprising, became a rallying cry for early abolitionists.[40] Her 1794 elegy for Marie Antoinette, while not an apology for monarchy, was a condemnation of the violence into which the Revolution had descended. The Latin and Greek pseudonyms she adopted in print – Constantina (constancy) and Philenia (lover of humankind) – signalled her humanitarianism and the neoclassical aesthetics that guided her work.

The discrete pearls she wears in Gilbert's portrait – socially conventional for a woman of her status – are the perfect subterfuge for a woman writer boldly staking her claim in the literary marketplace. Writing during the tumultuous years following American independence, Morton and a small number of female contemporaries were defining what it meant to live a life of intellectual inquiry and

Gilbert Stuart, *Sarah Wentworth Apthorp Morton*, 1802–3, oil on wood panel.

contribute to civic discourse.[41] Bolstered by Enlightenment ideals of women's moral and intellectual parity with men, Morton could apply the same artistic control to the short poem 'Memento: For My Infant, Who Lived but Eighteen Hours' as she could to her military epic 'Beacon Hill' (1797), dedicated to the soldiers who fought under Washington.[42] In her introduction to the latter, she argued that 'an author should be considered of no sex.'[43]

The polished persona presented by Stuart Gilbert is, rather, of both genders. Morton was a renowned beauty who balanced her social obligations with her creative work. She admitted 'Beacon Hill' had been interrupted 'When the season recalled me to the busy world ... and the appropriate occupations of my sex and station prevented even a thought of its continuance.'[44] She did, of course, prevail and the poem was sufficiently admired that George Washington preserved two presentation copies in his library at Mount Vernon.[45] As an elite society hostess, Morton inhabited the world of pearls and the drawing-room; as a writer with access to a blossoming print culture, she could help forge the new Republic's identity and secure opportunities for both men and women.

Around the time of the portrait's completion, the Philadelphia journal *Port Folio* published a verse exchange between the artist and his subject. Gilbert Stuart's lines to Morton penetrated to the heart of Morton's fine balance between masculine and feminine, public and private, determination and restraint:

> Nor wonder, if, in tracing charms like thine,
> Thought and expression blend in rich design;
> 'Twas heaven itself that blended in thy face,
> The lines of Reason, with the lines of Grace.[46]

KINDNESS: PEARLS EVERYWHERE

The fashion designer Gabrielle 'Coco' Chanel (1883–1971) has been called many things: daring, shrewd, defiant, subversive, single-minded, cunning, but seldom kind. From her birth in a charity hospital to unmarried itinerant traders, through an austere childhood at a convent orphanage, she rose to become a renowned couturière, a legend whose brand remains among the world's premier fashion houses.[47] 'Kind' seems an unlikely description of a woman who, on her scramble to dominance, denied the embarrassing existence of several siblings, underpaid her models ('They're beautiful girls. Let them take lovers') and had an affair with a German officer during the Nazi occupation of Paris.[48] Given the privations of her youth, it was hardly surprising that she sometimes seemed motivated by envy – kindness's opposing sin. But, if it was a kindness to free women from corsets, hobble skirts and stifling social norms, and liberate them to swim in the ocean, ride astride horses like men and possess their own bodies, then Coco Chanel was kind.

She effected her revolution through fashion, borrowing the shapes and fabrics of menswear, adopting simple, figure-skimming silhouettes, utilitarian materials like machine-knit jersey, and working-class clothing types like fishermen's tunics, sailor tops and mechanics' dungarees. She also achieved it through pearls. Chanel is the reason why pearls look good with anything, from chunky jumpers to silk shirts to bra tops. She trained our minds and eyes to obliterate the distinction between classic and contemporary, day-time and evening, working and leisure, highbrow and popular, male and female. She wore mismatched black and white pearl earrings and practised the difficult kindness of the iconoclast.

So many portraits of Chanel show her in pearls – slung over a tweed jacket, falling in waves over a little black dress, studding a straw boater, paired with a satin bow and cigarette. Pearls were as totemic as her lucky number, five. She had lavish collections of gems, gifted to her by wealthy lovers such as Grand Duke Dmitri Pavlovich, cousin to the Russian Czar, and the Duke of Westminster, who presented her with ropes of pearls every birthday. (When he gave her one to apologise for flirting with a younger woman, she dropped it disdainfully over the side of his yacht.)

She brazenly mixed real pearls with costume jewellery that she designed herself. Her Russian Romanov jewels inspired a theatrical

Unknown artist, Gabrielle Bonheur 'Coco' Chanel,
published in *The Bystander*, 10 April 1929.

line of gilt chains with baroque pearls and enamel crosses. She wore
them with metres of artificial pearls too bling to fool anyone, a sly wink
at bejewelled queens like Elizabeth I or Catherine de' Medici. During
her childhood, the only pearls that existed were natural ones, worn
with evening gowns by the aristocracy. A steady market for fakes had
existed for centuries, the most convincing created from glass beads
coated with *essence d'Orient* – fish scales suspended in oil. By the time

Sister Tilda NexTime of the Sisters of Perpetual Indulgence, self-portrait, 2019.

of Chanel's childhood, fake pearls were for the stage or the aspirational bourgeoisie, intended to mimic the real thing, while giving themselves away by their theatrical or middle-class wearers. Chanel's faux pearls, nestled between glass 'emeralds' and crystals, were unabashedly costume. Wearing them beside the real thing was a way of thumbing her nose at the society matrons who dismissed her as 'trade', or the aristocrats who gave her rubies but never a wedding ring.

Wearing pearls over jersey snubbed the convention of saving jewellery for evening. She went further. If pearls could be worn not just to the opera but for shopping and the racecourse, why not to the beach? Biarritz as well as the ballroom. She remarked to her biographer that she'd noticed a group of American girls swimming in Venice's lagoon and thought, 'How much more beautiful these young women would be . . . if they had dipped their pearls into the waves, into the sea from which they first came; and how brightly their jewellery would glitter if worn on a skin bronzed by the sun.'[49]

Chanel's costume jewellery revelled in artifice, but neither it nor her sailor blouses and swimming costumes were cheap. That hardly mattered; then, as now, couture was copied. There were, in fact, copy firms specializing in mass-producing new fashions, and, unlike other couturiers, Chanel was flattered that her designs were so popular. High fashion morphed to high street. Androgyny became chic, and just as jodhpurs, trousers, tweed jackets and cardigans became acceptable for everyone, pearls (or fakes) became ubiquitous. That pearls are now for everyone, joyously gender fluid, sported by style icons like Harry Styles and the Jonas Brothers as well as by their grandmothers, owes much to the kindness of Chanel.

GENEROSITY: GIVING AND RECEIVING IN MUGHAL INDIA

Mumtaz Mahal's husband and their eldest son admire jewels together in an intimate moment preserved in one of the most sumptuous albums to survive from Mughal India. Eleven years later, this child will lose his mother. The husband, in his grief, will commission an enduring symbol of devotion, the Taj Mahal. The son will grow up to be an art patron, philosopher and mystic, beloved for promoting interfaith harmony. Although favoured to succeed to the Mughal throne, at 44 years of age, he will be assassinated by his power-hungry younger brother.

In 1620, though, all this lies in the future. At the centre of the image, the young father and his five-year-old sit together in a garden. Birds of paradise, cranes, partridges, peacocks, roses and poppies decorate the image borders, as they must have decorated this refined and privileged life. The short sword tucked into the child's *patka*, or waist sash, is just for show. More important is the purple iris he holds up for his father, or maybe it's an ornament for his turban, a birthday present for a favourite son. At his birth, his grandfather

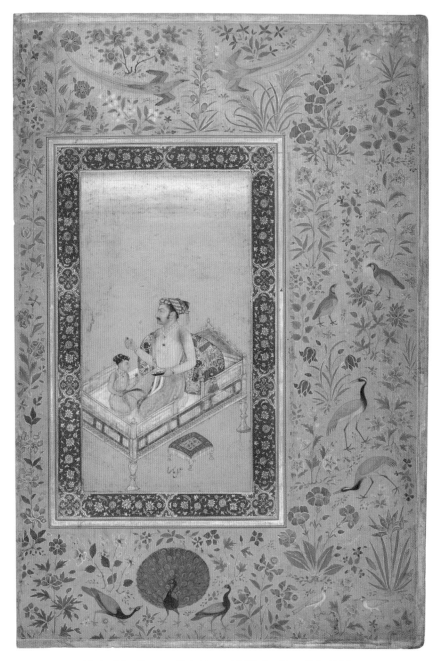

Nanha, 'The Emperor Shah Jahan with His Son Dara Shikoh', folio
from the Shah Jahan Album, *c.* 1620, ink, opaque watercolour and gold
on paper; margins: gold and opaque watercolour on dyed paper.

noted Sagittarius was in the ascendant – an auspicious sign that he hoped to mean the child would bring his family good fortune.[50]

The painted moment distils the pair's pleasure over a plate of jewels, among the most gifted items in Mughal India, the most popular of which were pearls. The father, Shah Jahan (King of the World, r. 1628–58), and his son Dara Shikoh drip with the white gems. The child alone wears five strands around his neck and one looping over his turban. He has pearl bracelets around each wrist and pearl earrings as large as the iris of his eye. Some might well have been gifts from father to the apple of his eye. Shah Jahan, too, is decked out in pearls, rubies and emeralds – gems that he often received as presents, and his favourites for dispersing to family and friends.

Around the time of this portrait, Mughal artists were achieving a realism unprecedented in the Islamic world. This naturalism allows us to identify the sitters as well as the species of flowers and animals that animate these 'family albums' of the Mughal court.[51] Given these artists' adherence to authenticity, it seems probable that some of the pearls they depicted were quite large in real life. The one dangling from Shah Jahan's ear is the size of a thumbnail. Measured against his iris, several of his pearls would be well over a centimetre in diameter. Although Mughal India imported pearls from the Persian Gulf, the most common species there, *Pinctada radiata*, doesn't produce large stones. More likely, these extravagant baubles came from *Pinctada maxima*, traded along well-established routes between India and Southeast Asia.[52]

To hoard is human, to give is divine, a reflection of Allah's generosity to his children. The Prophet Muhammad, who happily gave and received, taught that generosity cultivates love and removes rancour from the heart. Giving is a way to share the nectar of the created world and relish its wonder. But gifts also lubricate social relations, cement political alliances, offer tribute and reward loyalty. As much about diplomacy as delight and devotion, generosity was perfected at the Mughal court, and pearls played an outsized role.[53]

The Mughal Empire, ruled by Shah Jahan during a period of titanic wealth and power, lasted three centuries, but its gift-giving conventions reached far back through courtly tradition in the Islamic world. The world's sweetness was offered to allies, trade partners, benefactors and rulers in the form of costly fragrances and spice, furs, velvets, silks woven with gold, camels and horses, and calligraphy that turned each word of scripture into an embodiment of God's beauty. Meticulous records document presents in pre- and early Islamic periods, such as a gift offered by the king of India to the Persian king Khosrow I (d. 579), of aloeswood incense 'which melts in fire like wax', a ruby goblet filled with large pearls, and a letter written in mother-of-pearl on fragrant camphor bark.[54] A later Indian king gave one of his governors a disturbing (to us, at least) toy camel on silver wheels, whose udders could be milked for small pearls and throat slit for rubies.[55] A regal marriage was marked by the bride's grandmother showering pearls on female guests; the bride received a golden tray rattling with a thousand pearls.[56]

When Shah Jahan presented exceptionally fine pearls to his children, he was following the precedent of centuries. By the time of his reign, spates of pearls, rubies, rosy spinels and emeralds washed through the court. Accounts of the quantities of gems flowing between family members, politicians and foreign dignitaries suggest that nothing rested for long, locked in its wooden cabinet on a velvet-lined tray. Pearls were given and received on birthdays – most spectacularly on the emperor's birthday, when he was weighed on a huge jewelled scale against gold, silver and gems. The heavier the emperor, the more numerous the gifts distributed to those whose happiness hung, quite literally, in the balance.[57]

Gifts of pearls marked achievements as diverse as military victories, the concoction of a new rose perfume, a baby's birth, a child's memorization of the Quran.[58] According to the French gem trader Jean-Baptiste Tavernier, who witnessed the birthday weighing of the next emperor, the royal family had an ingenious strategy for ensuring the best stones stayed in the royal treasury. Courtiers, obliged to present their emperor with expensive gifts, bought jewels from the treasury so they could gift them back – a wily route for the emperor to benefit twice.[59]

Chitarman, 'Shah Jahan on a Terrace, Holding a Pendant Set with His Portrait',
folio from the *Shah Jahan Album*, 1627–8, ink, opaque watercolour and gold on paper.

Forty-one years after Shah Jahan admired jewels with his adored son Dara Shikoh, another pearl made a sad coda to their story. In April 1661 the new emperor Alamgir (World Seizer, r. 1658–1707), the third son of Shah Jahan and Mumtaz Mahal, received a distinguished visitor. The Iranian ambassador arrived bearing congratulations on Alamgir's accession to the throne. As customary, he also brought gifts – camels, Arabian horses and a single large and lustrous pearl valued at the price of a fine mansion.[60] He was received in a white marble hall in the palace at Shahjahanabad, the city built just a few years earlier by the emperor's father. The occasion's decorum belied the bloody nature of the succession. Shah Jahan was still alive, languishing in Agra's Red Fort, imprisoned there by Alamgir. The favourite, Dara Shikoh, had already been murdered, and the new Mughal emperor was cleansing his circle of siblings and rivals.[61]

The gift of an incomparable pearl conveyed time had moved on. Emperor Alamgir now sat on his father's jewel-crusted 'Peacock Throne'. He accepted booty from every corner of his empire. But with regards to his murdered family, generosity engendered no love.

DILIGENCE: TRACKING DOWN A PEARL THIEF

Pearls – small, transportable, impossible-to-trace stores of wealth – have made irresistible targets for thieves. Implanting the most valuable with identifying microchips is now theoretically possible, but for most of their history, stolen pearls became anonymous the moment they were detached from a strand. Some heists are legendary, such as the notorious 'great pearl robbery' of 1913, where a necklace worth £9 million was stolen from a prominent London dealer. This 'Mona Lisa of Pearls' was intercepted in the mail by a debonair crook who replaced the necklace with eleven lumps of sugar.[62] Some heists are fictional, such as Marlene Dietrich's swiping of a fabulous pearl necklace in the 1936 romcom *Desire*. Dietrich's better angels convince her to return the strand, but most cases of recovered pearls rely on dogged diligence and old-fashioned footwork.

A robbery in Shanghai in 1918 illuminates policing at its least slothful. Just after 8 a.m. on 14 December 1918, the staff of Mikimoto's pearl shop on Shanghai's prestigious Nanking (Nanjing) Road arrived to discover the shop had been burgled.[63] Always sensitive to the optics of business, in 1916 Mikimoto had chosen an opulent venue to market his new cultured pearls to the fashion-forward Shanghainese. His shop stood on the smartest stretch of Nanjing Road, beside the new department stores Wing On and Sincere, and stocked not only his miraculous cultured pearls but gold, platinum and precious gems.

In 1842 Shanghai had become a Treaty Port, allowing foreign residents to live under British law and administration. It was, then, a British officer from the Shanghai Municipal Police, which had been set up in 1854 to govern the Shanghai International Settlement, who was called to investigate the pearl heist. When Inspector James Cruickshank arrived at Mikimoto around 8.30 a.m., the broad, paved boulevard was already seething with traffic. Rickshaw carriers dodged trams clanking westwards into the French Concession, and hooded black motorcars whisked Europeans to their banks, insurance offices and consulates on the Bund – the imposing stretch of foreign buildings along the Huangpu River.

As commonplace as road accidents, burglaries were routine, even mundane, for a seasoned inspector like Cruickshank. He'd served for fourteen years with the Shanghai Municipal Police, and during those years, he'd encountered everything: gang murders, drug raids, extortion, all the shades of racketeering that fed on the immense new wealth that gushed through the city like the wide, rank waters of the Huangpu in which coffins bobbed and sometimes snagged on wharves stretching like fingers into the river. Before Shanghai, he'd served in British South Rhodesia.[64] Born in Aberdeen, he had the Scots' restless nature and feet that itched to leave northerly winds and granite buildings behind. Eight years after the Mikimoto burglary, he would retire, amply decorated for his bravery under fire. A photograph taken around that time shows a solid, resigned-looking man in his middle years, four medals pinned to his white jacket, his narrow black satin tie smoothly knotted at his throat.[65] His remaining

M

大志 *Tse-dah*

3E Peking Rd. Tel 893
Tel Add : Middy

Middleton & Co.
(Shanghai), Ld.

Merchants, Commission and
Manufacturers' Agents

Middleton, W. B. O.,
Mging Direc.
Andersen, A. E., signs *p.p.*
Standley, W. A.
Murphine, S.
Greiner, Miss

Agents—
The A. Butler Cement Tile
Works, Ld.

———

號笠三 *San-lih-ho*

43 Bub. Well Rd. Tel 3362

Mikasa & Co.
Japanese Silk and Cotton
Goods, etc.
(See Advertisement, page 7)

———

行珠珍本木御 *Yu-mo-pen-chen-tse-hong*

31 Nanking Rd. Tel 3094
Tel Add : Pearlmiki

Mikimoto, K.
Dealer in Pearls and Jewels

Mikimoto, K. (Tokio)
Kambe, K., *Mgr.*
Moh, F. D.
Oda, Y.

———

報論評氏勒密 *Mee-lard-sze-pin-lun-pao*

113 Avenue Edward VII
Tel 4747 Tel Add : Millard

Millard's Review

Publishers : Millard
Publishing Co., Inc.
Millard, Thomas E.,
Editor and Publisher
Powell, J. B., *Financial Edr.*
Missemer, Geo. W., *Sub-Edr.*
Chang Sing-hai, Business
Dept.
Zung Nyok-yien, ,,
Wong Yeh-san, ,,

———

廠木昌銘 *Min-chong-mo-chon*

459 Manila Road, near Great
Western Road

Min Chong & Co.
Builders, Painters, Reinforc-
ed Concrete and General
Contractors

Downg, M. C., *Gen. Mgr.*
Downg, F. L., *Asst.*
Wong, T. J., *Accnt.*
Tsu, Z. L., *Clerk*

MISSIONS

MISSIONS
(See also Churches)

會禮浸國美 *Mei-kwok-tsin-lee-wei*

9 Hankow Road
Tel Add : Baptisma

American Baptist
(Northern) Mission

Proctor, J. T., D.D. *Sec.*
Stafford, R. D., *Treas.*
Dahl, Miss L. J., *Asst.*

———

會差禮浸國美 *Mei-kwok-tsin-lee-char-wei*

177 North Szechuen Road

American Baptist
(Southern) Mission

———

9 Hankow Road

American Baptist Foreign
Mission Society

Proctor, J. T., and wife
Stafford, R. D., and wife
White, F. J., and wife
Huntley, Dr. G. A., and wife
Mabee, F. C., and wife (ab)
Kelhofer, E., and wife
Bromley, C. L., and wife
Kulp, D. H. and wife
Hanson, V., and wife
Dahl, Miss L. J.
Reeder, R.

———

會經聖國美大 *Ta-mei-kwok-sing-ching-way*

53 Szechuen Road
Tel Add : Bibles

American Bible Society

(China Agency)
Hykes, Rev. J. R., D.D., *Agent*
and wife
Cameron, Rev. W. M., and
wife
Silva, J. M. B. da
Ferris, Miss R. S.
Taylor, Miss J.

———

會公聖 *Sing-kung-way*

Office : 6a Seward Road
Tel. 1868
Tel Add : Jessfield

American Church Mission

Smalley, S. E., and wife
Wilner, R. F., and wife
St. Luke's Hospital
12 Seward Road
Tucker, A.W., M.D., and wife
Morris, H.H., M.D., and wife
Bender, Miss M. E.
Chisholm, Miss E. S.
5 Yu Yuen Road
Merrins, E. M., M.D., and
1 Avenue Road [wife
McRae, Rev. C. F., and wife
2 Avenue Road
Reid, Miss S. H.
Boone, Miss A. A.
Cartwright, Miss E., M.A.
St. Elizabeth's Hospital
3 & 4 Avenue Road
Fullerton, Miss E.C., M.D.

Agents: ANDREWS & GEORGE

Entry for Mikimoto K., 31 Nanking Road, Shanghai,
The North-China Desk Hong List, July 1917.

dark hair is clipped close to his skull. Perhaps because of his baldness, his ears seem unusually large. He looks like a man trained to pay close attention.

Given the store's fame, and Concession-era Shanghai's extreme inequities in wealth, Cruickshank wasn't surprised Mikimoto had been targeted. He was surprised that the thieves hadn't bothered to pick the front door's Yale lock but had simply smashed open the door. He instructed his men to dust for fingerprints and went inside to survey the damage. The thieves had gone straight for the safe. A hole had been drilled on its left side, through which they'd accessed the keyhole. They'd wrenched out the brass fittings and tossed them to the floor, making it easier to pick the lock. Cruickshank instructed his men to pick up the metal debris and crumpled paper and book them as evidence. Surveying the damage, he could see that the thieves were skilled, but they were careless and impatient. Nanjing Road was electric at night, and it would have been far less risky to pick the entrance door's Yale lock rather than muscle through it. It was clear, too, that they hadn't worn gloves. Greasy fingerprints smeared the safe. If Cruickshank was lucky, they'd be able to lift clear impressions of the whorls, loops and arches that could tip them off to the culprit more eloquently than any informer.

The store manager, Kōkichi Kanbe, called them to a back room, where a roll-top desk containing precious unset gems had been ransacked. All in all, the thieves had made off with a fortune in pearls, gold, jade and other jewels. Unperturbed, insured and recognizing a publicity opportunity, Mikimoto bought an advertisement in the Shanghai papers, promising a generous reward for information leading to his pearls' return. But before anyone could come forward, Cruikshank, having first applied for a search warrant from the Russian Consulate, made a raid on a Russian boarding house in the northern district of Wayside. Two residents, Adam Jembitsky and Adolph Zilberstein, had just sat down to tea and cakes when Cruickshank and his men strode into their apartment. Protesting their innocence, the two men were arrested and booked at nearby Hongkou Police Station.

Wayside was an underbelly of the glittering metropolis that Cruickshank knew all too well. At the time of the Mikimoto robbery, Shanghai newspapers would claim it was where 'Russians of the suspicious class reside'. It was, in fact, where the dispossessed and desperate resided. In 1918 there were almost 1,000 Russians in Shanghai. Some were businessmen, but many were refugees who had fled the Russian Revolution. They came to be known as the White Russians to distinguish them from communist sympathizers. Typically upper-class professionals and intellectuals, they had a hard time adjusting to life in Shanghai, where they didn't speak the language and couldn't easily find work. Men especially found themselves unemployable, and alcoholism was rife. The fact they were there at all was only due to China not requiring entry papers and visas, a unique situation that two decades later would make Shanghai an ark for Jews fleeing the Holocaust. Like the Russians, they would make a temporary home in Wayside.[66]

At Hongkou Police Station, Cruickshank's diligent detective work paid dividends. The prints his officers had lifted from the safe were a match for the three, long, tapering fingers of Jembitsky's left hand. He had, the detective theorized, steadied himself against the safe as he drilled a hole in its side. A search of the boarding house turned up a chisel belonging to Zilberstein, which matched marks gouged into the woodwork of the manager's looted rolltop desk. But as for the pearls themselves, there was no trace.

Three months later, Sergeant Ferguson was enjoying his customary dawn walk through the Shanghai Public Gardens after his night on duty. If not quite as veteran as Cruickshank, he too was a seasoned policeman, and so when he saw someone scratching in a flowerbed at dawn, he was alert to trouble. The man appeared to retrieve something from the soil, then walked north towards Suzhou Creek. While digging in a flowerbed might not be illegal, it was at least suspicious. Despite being off-duty, Ferguson shadowed his prey until the man, realizing he was being followed, raced towards a bridge and hurled two objects into the water. As Ferguson caught up with him, the officer caught his second lucky break. Sergeant Constable

Chegwidden of the River Police happened to be patrolling the creek in his boat. Ferguson grabbed his suspect and asked Chegwidden to dredge the creek. He soon retrieved two sodden but intact objects.

A few days later, Mikimoto's staff was cleaning the pearls that had been wrapped in newspaper and stuffed into two cushion covers. Despite their burial and submersion in Suzhou Creek, they were intact, and soon the jewellers had them gleaming. It was a double boon for Mikimoto, who recovered his merchandise and attracted a fresh round of publicity. For the Pearl King, at least, being a burglary victim was good for business.

Meanwhile, the man apprehended by Sergeant Ferguson was processed through Shanghai's Central Police Station and found to be none other than Adolph Zilberstein, one of the pair arrested the previous December in Wayside. He and his partner Jembitsky had been taken to the Russian Consulate and shipped north to Vladivostok for trial. Jembitsky's fingerprints were enough to convict him, but as there was not enough evidence against Zilberstein, the latter was released. On the night of Sunday, 2 March, he boarded a train back to Shanghai to retrieve the stolen jewels. By Monday morning, after the sheer bad luck of being caught in the act by Sergeant Ferguson, he was back in custody. Within days, the Russian Consulate arranged his return to stand trial in Vladivostok.

It was a lucky accident that Sergeant Ferguson caught Zilberstein red-handed. But it was diligence that had tied the two men to the original theft. The previous December, Cruickshank's men lifted perfect fingerprints from Mikimoto's premises. But it wasn't the pattern of whorls and arches that led the police straight to the Russian boarding house in Wayside. At that point, they had nothing to match the prints to. It was only after Jembitsky was fingerprinted at the station that his guilt was apparent. When Cruickshank's men picked up the brass fittings that had been prised from Mikimoto's safe, they also picked up a piece of crumpled paper. It was a page of a Russian newspaper Jembitsky had used to wrap his tools, printed in Harbin, a northern Chinese city that had absorbed 100,000 Russian refugees. The newspaper suggested that the thieves were probably Russian, and

Cruickshank only had to make a few inquiries among his Russian contacts in Wayside to discover that one Adam Jembitsky used to work in a safe factory.

SIX

❀

Embodied

Sometimes pearls are known to us only through art, captured in a painted portrait, or fixed in a photograph. Because of their relative softness and fragility, they are more easily damaged than harder gems. Often small, they are apt to be misplaced, skitter under furniture or break loose from thread. The most innocent gestures – a misdirected spray of perfume – discolour them. Perspiration dulls their nacre. Worn too seldom, stored in a warm, dry place, their nacre cracks. Claude Arpels, who was sent on 'jewel safaris' for Van Cleef & Arpels, recalls appraising a cache of pearls from the estate of an Indian maharajah. In an episode worthy of religious allegory (or Disney's *Pirates of the Caribbean*), when he reached into the brimming coffer of gems, its contents crumbled to white powder. Imprisoned for years in an airless vault, the pearls 'were dehydrated and dead as ancient bones'.[1]

Loved too much, locked away, they desiccate. Or they are spirited out of sight. History is full of legendary pearls rumoured to have belonged to Marie Antoinette, Catherine the Great, Empress Eugénie. The more famous and the more desired, the more likely they are to drop out of history and disappear into a private collection, perhaps resurfacing decades or centuries later. Often they are best preserved by artists rather than their wearers.

Pearls are grown by living creatures and, because of their surface structure, gleam with light. But how does an artist bring a pearl to life?

What pigments duplicate pearls' luminescence, and what medium embodies the lustre of nacre? What captures pearls' intimacy, glowing warm against skin, grazing a wrist bone, gently tugging an earlobe or reflecting a collar? How does a portrait convey not just our status but the complexities of our millennia-long relationship with pearls?

PLAYING THE LONG GAME

Among a remarkable group of portraits from Roman Egypt is one of a woman whose name, Isidora, is inscribed in Greek. Her dark hair is coiled on top of her head and adorned with a gold wreath. Tight curls frame her face and full brows arch above clear brown eyes. Only the skin of her neck betrays the beginnings of middle age. Her expensively dyed lilac robe sets off her glowing pale gold skin and rose-pink lips. Her robe is the same colour as an enormous amethyst pendant, attached to a necklace of gold and green gems. She is well-groomed, glowing with health, affluent. In addition to her amethyst pendant, she wears a necklace of heavy gold beads and another with gold, green gemstones and pearls. Large pearls dangle from extravagant gold earrings.

These portraits are the painted jewellery boxes of Roman Egypt. Their subjects drip gems. One was so decked out that Flinders Petrie, the Egyptologist who discovered her in 1911, dubbed her 'the jewellery girl'.[2] She has pendant pearl earrings like Isidora's, a choker with pearls and green stones, and a pearl pin in her braided hair.

Just over a thousand of these portraits survive, dating from the first century CE to the mid-third century, when Egypt was a province of Rome. A surprising number of them portray pearls.[3] Most depict well-presented people in their twenties, thirties and forties, though some are children and adolescent males with the first wisps of facial hair. Their dress and jewellery suggest that these sitters are in the prime of life and peak of their economic and social success. When

Mummy Portrait of a Woman, Romano-Egyptian (Egypt), 100–110 CE, encaustic on linden wood, gilt, linen.

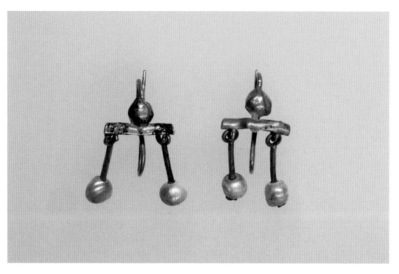

Earrings, Roman, 1st century CE, gold, pearl.

Flinders Petrie began assembling a portfolio of portraits excavated from the vast necropolis of Hawara in the Fayum Oasis, he was struck by their abundant jewellery. There were surfeits of pearls, emeralds, rubies and gold. Winking discs of gold weighed down ear lobes. Necks were draped with pearls, chunky gold chains and grape-sized emeralds. But despite the bling, these accessories resembled actual Roman jewellery preserved in museum collections. Petrie realized that matching the painted jewellery with surviving examples could help date the portraits. The bar and pendant pearl earrings worn by Isidora were of a type that had attracted the ire of Pliny the Elder, who spluttered against the

> wearisome refinements which have been devised by profuse-
> ness and prodigality; for after inventing these earrings, they
> have given them the name of 'crotalia' or castanet pendants,
> as though quite delighted even with the rattling of the pearls
> as they knock against each other.[4]

Many of these crotalia were found at Pompeii and Herculaneum. Some, with pearls dangling from plaited gold, had been found in

Egypt in the region of Petrie's excavations.[5] Isidora's *crotalia*, with four pendant pearls rather than the more usual two or three, are wildly extravagant.

The portraits' subjects were the elite citizens of Roman Egypt. Many resided in the wealthy towns of the Fayum basin, an oasis in the Nile Valley whose fertility made it the breadbasket of the Roman Empire.[6] Greeks had emigrated to the region following the invasion of Egypt by Alexander the Great in 332 BCE. Under Rome from 30 BCE, the Fayum was a cultural oasis irrigated by Hellenistic, Roman and native Egyptian influences. Portraits are inscribed with Greek, Egyptian and Latin names. Their owners would have had access to luxury goods such as costly dyes and gemstones that travelled along the trade networks of the empire, allowing them to keep abreast of metropolitan Roman fashion. Bordered by the Mediterranean and the Red Sea, Egypt was at the crossroads of trading routes that reached into Europe, Arabia and India.

The pearls depicted on these portraits could have come from the Indian Ocean, the Persian Gulf or much closer to home, from fisheries in the Red Sea.[7] In the Pharaonic period, there is little evidence that pearls were particularly valued.[8] They became more popular under the Ptolemies, the Graeco-Macedonian last dynasty of Egypt. Under the Romans, pearls were a fashion rage. The difficulties of keeping up with demand led to rampant smuggling as well as recipes for fakes.[9]

The materials for the portraits, too, relied on Roman trade networks. In many cases, the timber used for the panels was linden (*Tilia*), imported from Europe. This lightweight, easily worked wood had a fine, even grain, no discernible knots and insecticidal properties that made it a popular choice for the artists and workshops who produced these portraits.[10] The pigments were sourced from across the empire. The red lead (Pb_3O_4) which appears on Isidora, was a toxic by-product of Roman silver mining at the colossal Rio Tinto mines in Andalusia. These mines might also have been the source of a yellow pigment, jarosite.[11] Lead white from the mineral hydrocerussite was a relatively new pigment that entered Egypt during the Graeco-Roman period. It, too, was a mining by-product and may have originally come to Egypt from the Athenian silver mines at Laurion.[12] Lead white appears extensively on these portraits. Non-destructive imaging such as X-radiography and X-ray fluorescence mapping shows how commonly this pigment was used as a ground and a base for dewy flesh tones. And for pearls.

This opaque, warm white, though toxic, was perfect for pearls. Under raking light, Isidora's pearls pop in three dimensions. This effect is partly dependent on the portrait's medium, encaustic, in which beeswax is mixed with pigments and worked with great skill while hot. Isidora's heavy impasto pearls have physical heft. Her gold jewellery is gilded with wafers of actual gold, probably from Nubia in southern Egypt, long a rich source of gold deposits. Her face springs alive through the visible tool marks of an artist who sculpted paint around her eye sockets, around her nostrils, and into the groove above her lips. The product of living creatures, the pigmented beeswax catches the light and animates her personality.

These portraits have been admired for their naturalism since their discovery. Their sitters might just have paused talking and turned their attention towards us. But these were portraits of the deceased and most likely painted post-mortem.[13] Although hailed as the ancestors of the portraiture tradition in the West, they were embedded in Egyptian mortuary practices that reached back to around 2600 BCE and the beginnings of mummification rituals. These paintings, whose fresh spontaneity has been compared to twentieth-century masters such as Henri Matisse and André Derain, were tightly bound into their deceased owners' mummy wrappings. Most have become detached from their mummies, with only about one hundred surviving intact. Although Isidora's portrait at some point separated from her mummy, enough survives of the case to reveal how these portraits originally appeared. Isidora was one of a tiny subset of about twenty mummies termed 'red shroud mummies', as their linen wrappings were painted with red lead and then decorated.[14]

But it was in the service of life, not death, that the deceased were provisioned with portraits capturing them at their finest and such care was lavished on their bodies. Egyptians loved life so much they rebelled against death. Isidora doesn't wear her pearls in the hopes of being remembered as a person of high status. She wants to live on, not in people's thoughts, but in the company of gods. Every service performed for the dead was to ensure the immortality of a transfigured being. Mummification – the careful evisceration and desiccation

Mummy Portrait of a Woman, detail of pearl earring
in raking light taken by conservator Marie Svoboda.

of the corpse, its preparation with tree resins, oils and perfumes, and its wrapping and decorating – freed the human to survive death as a perfect being.[15] The portraits inserted into the wrappings did record the distinctive features of individual subjects, but at their most vibrant and attractive.[16] The mummies were the perfect cocoon for the soul to return to each night. Red lead gave the privileged few red shroud mummies like Isidora an extended warranty on immortality, as the pigment was moisture-proof and a toxic pesticide.

Isidora, like other elite women of Roman Egypt, wore her pearls to traverse eternity. In her life, the strands of Egyptian, Greek and Roman culture entwined like the tributaries of the Nile. Deities were as fluid as fashions, and fusions between the Egyptian Isis and the Graeco-Roman Venus Aphrodite were common. In 79 CE, when Isidora might have been a teenager, Mount Vesuvius erupted, burying Herculaneum and Pompeii. Both Isis and Venus were worshipped here.[17] A painting of Venus in a pearl choker and pearl earrings like Isidora's was excavated from the ruins of the sumptuous Pompeiian house of Marcus Fabius Rufus. Pearls were a perfect tribute for the goddess Venus, who comfortably morphed into Isis, who was identified with female mummies.[18] Twenty years after the painting of Venus disappeared under rubble and ash, Isidora's pearls were a faultless accessory for her meeting with the gods.

PEARLS ON BLACK BODIES

Cradled by artificial lakes and the humanist ambitions of its rulers, the Renaissance city-state of Mantua in northern Italy was a fertile environment for artists and scholars. The court painter Andrea Mantegna spent over four decades there in the service of the governing Gonzaga family, nourished by the intellectual vitality of the court, the family's sumptuous collections of art and antiquities, and the moisture-soaked landscape. Around 1500 he completed an unusual

Unknown artist, statue of Venus, Roman,
2nd century CE, partially gilded bronze, gold, pearl, glass.

devotional painting. Its scale and its subject-matter – a close-up portrait of the three kings who had come to worship the newborn Christ in Bethlehem – suggest that Mantegna may have created it for a domestic space or small private chapel, probably for a member of the Gonzaga family.[19]

By this date, the subject of the magi was a familiar one in Christian art. What was less familiar, particularly in Italy, was the close-up depiction of the youngest of the kings as a black African. An extraordinary piece of jewellery loops from his pierced ears, made from pearls strung on a gold chain, interspersed with red, blue and yellow gems. It is unmatched in the Renaissance, and it would take the imagination of Cartier over four centuries later to create a pearl necklace that is attached to the ears.[20]

The magi are mentioned only in the Gospel of Saint Matthew, which claims they came from 'the East', with gifts of gold, frankin-cense and myrrh. Matthew uses the Greek word *magos*, probably to refer to the learned Persian-Anatolian priests adept in interpreting dreams and visions.[21] In early depictions, they are Persians. The Tunisian-born Early Christian theologian Tertullian connected Matthew's narrative back to passages in the Old Testament, which prophesied that *kings* from distant shores would bring tribute to a royal son.[22] Although the writer of Matthew didn't specify how many magi journeyed to Bethlehem, the different types of gifts suggested three, and it became a convention to depict them as the Three Ages of Man. They attained cult status, with religious centres vying to display relics such as their bones. A popular *History of the Three Kings*, compiled in the fourteenth century by John of Hildesheim, claimed that one of the kings was a black Ethiopian.[23]

Northern European artists were quicker than their southern counterparts to depict a black magus. Usually, he was identified as Balthazar, the youngest of the kings, to convey the idea of Africa being more recently introduced to Christianity than Europe and West Asia. Balthazar became a metaphor for the spread of faith, but as a black king, he is a paradox. His appearance in art coincides with the rise of the slave trade from the mid-fifteenth century and cargoes of

Andrea Mantegna, *Adoration of the Magi*, c. 1495–1505, distemper on linen.

enslaved people imported into southern Europe from West Africa. In northern Europe, where slavery was illegal, King Balthazar was most likely to be depicted as black.[24] In southern Europe, this was a much later pictorial development. The enslavement of white peoples was entrenched in late medieval Italy; from the mid-fifteenth century they were joined by an influx of black people from sub-Saharan Africa. Perhaps this swayed how Italian Renaissance artists depicted black Africans. Where black men are included in Italian narratives of the Adoration, they are most often servants or attendants of white kings.[25]

But in addition to enslaved people, there were other ways in which Europeans became familiar with black Africans. In Renaissance Italy, slavery was not a life sentence. The enslaved could be freed to work, most famously, as gondoliers in Venice. Their children could become lawyers, teachers, artists and clergy.[26] There were, of course, people who had never been enslaved. During the fifteenth century, diplomatic, commercial and political ties strengthened between European and African countries. Ethiopia, which had a long Christian tradition, sent scholars and pilgrims to Rome, where there was a dedicated residence for Ethiopian visitors.[27] European courts hosted high-status African ambassadors and their retinues.

One of Mantegna's earliest works for his Gonzaga patrons was a 1460 *Adoration of the Magi*, now in the Uffizi. A black king, accompanied by several black attendants, kneels with his gift. It was the first time a major Italian artist had depicted a black African magus. The prolific turbans and camels owe more to European imagination than to knowledge of Africa. Still, four decades later, Mantegna had plenty of opportunities to observe people of African descent at the Gonzaga court. Isabella d'Este, who married into the clan in 1490,

Adriaen Hanneman, *Mary I Stuart with a Servant*, 1664, posthumous portrait, oil on canvas.

was especially drawn to black attendants. Shortly after she arrived in Mantua, she requested the purchase of a small girl 'as black as possible'.[28] Her appetites were not unusual in European courts, where young, black children were often treated as exotic playthings for their owners, luxury items considered much like rare animals for private collections or precious gems imported from the mines of Africa.

From around 1500, it became fashionable for artists to depict European aristocrats with their black attendants. The legal status of these people is often unclear. Although labelled too frequently as slaves, they are not necessarily enslaved and are often dressed in shimmering satins, with pearls in their ears and around their necks. Their subservient status, though, is entirely unambiguous. A 1682 portrait of Charles II's mistress, Louise de Kérouaille, Duchess of Portsmouth, depicts her with an expensively attired black child, arrayed in pearls and offering the duchess a shellful of pearls and a branch of red coral. The duchess looks squarely at the viewer; the child looks adoringly at the duchess. The white pearls gleaming against the child's dark skin are as much the sitter's possession as the child is. Coral from the Mediterranean, pearls from the Persian Gulf or the Americas, a child from Africa – there was nothing out of grasp for the king's favourite. Or for England's overseas ambitions.

Mantegna's Balthazar, though, was more likely inspired by an ambassador than a servant, and his gems are his own possessions. Under successive Gonzaga scions, the Mantuan court emerged as a significant player in European politics and culture, spinning a web of alliances with the ruling families of Europe. The Gonzaga court hosted foreign dignitaries, eminent scholars, artists, poets and musicians. While Mantegna was familiar with the court's black attendants, he also knew about and possibly met elite black visitors.[29]

While the family collected people, they were equally renowned for collecting art. Isabella d'Este avidly collected not just paintings but engraved gems, antique medals and cameos. She had a documented taste for pearls, bringing a sumptuous strand with her on her marriage to Francesco. Later, Titian painted her wearing a large pendant pearl earring and a pearl ornament in her headwear.[30] Gonzaga patronage

Rosalba Carriera, *Personification of Africa*, before 1720,
pastel on paper mounted on canvas.

ranged from the commissioning of palaces and monumental paintings
to the acquisition of small, exquisitely crafted objects.

Mantegna, too, had a reputation as a gem expert, an obsession
that tended to drain his funds.[31] Although his *Adoration of the Magi* is
a devotional painting, it is also a celebration of connoisseurship and
the precious materials that Mantegna was surrounded by in Mantua.
Mantegna is a master of hard surfaces, revelling in the gifts presented
to the infant Jesus. There is the veined orange/red translucency of
the incense burner capped with silver. Balthazar holds a sinuously
banded agate vessel. The early fifteenth-century blue-and-white

porcelain from the Jingdezhen kilns in China is more precious than the gold coins it contains, because it is rarer. A fitting gift for a king, the secret of its glassy translucency would tantalize Europeans for centuries. And there is Balthazar's fantastical necklace-cum-earrings, the coloured gems like candied-sugar confections, and the pearls hard, dense and granular. Balthazar's pearls were painted by an artist who rubbed pearls between his fingers, knew each one was unique, and enjoyed their slightly perceptible grittiness at his fingertips.

To discriminate between all these textures, Mantegna painted this *Adoration* in distemper, a medium he favoured for his devotional works. It was a demanding medium and an unusual choice for an Italian artist. Instead of using oil as a binder for pigments, distemper relies on animal glue derived from the skins of sheep or goats. Applied in thin, opaque washes, it remains water-soluble and can be carefully blended to produce luminous, meticulously refined surfaces.[32] It was Mantegna's solution for presenting the world of minerals and gems in all their unmuddied brilliance.

In European paintings of the magi, Balthazar would often be depicted with a pearl earring, an image fuelled by travellers' tales of Africans with extravagant jewellery and piercings. Pearls were not typically sourced from Africa though; until the sixteenth century, Europeans sourced them from the Persian Gulf or northern rivers.[33] But symbolically, a pearl was a vague nod to Balthazar's exoticism. Artistically, it was an opportunity to contrast a gleaming white gem against black skin. As the sixteenth century progressed, the connection between pearls and African skin strengthened most bitterly. Needing a source of exploitable labour for the pearl fisheries of the Caribbean, the Spanish administrators petitioned the Spanish crown in 1518 to provide enslaved divers from Africa. From its early years onwards, the New World's pearl fisheries depended on the transatlantic slave trade.[34]

Balthazar's gift to the Christ child of myrrh, a gum resin used for embalming, presaged pain on so many levels. But Mantegna's Balthazar exists in a time before black bodies were paid for with gold, or sometimes pearls, to live out bleak lives in the pearl fisheries of the Americas. His pearl jewellery, acquired from somewhere far from home, is the mark of a king who received as well as offered tribute.

GIRL WITH A LEAD WHITE EARRING

Of the 36 paintings attributed to Johannes Vermeer, eighteen depict pearls. They lie casually on tables, dangle from ears, are pinned onto hair and are tied around necks with ribbon. Vermeer is so closely associated with pearls that the adjective most commonly used to describe his light is 'pearly'.[35]

Pearls were a flexible symbol, as the literature of the period makes clear, and could reflect the virtue and vanity of a subject.[36] They could also point to the Dutch Republic's maritime empire and the country's prestige at the nexus of trading routes to the Caribbean, India, northern Europe, Southeast Asia and the world's booming precious commodity markets. They took their place beside emeralds and tobacco, rubies and paintings, ermine and parrots. Several of Vermeer's pearl-decked women pose in front of maps or maritime paintings, trumpeting this achievement. More subtle, though, is the way the boundaries between domesticity and commerce collapse in these paintings. Maps bring the outside world inside, but so too could women. Women bought, sold and pawned pearls, and on occasion, drilled them, as well as wore them.[37] Neither pearls nor women are simple accessories.

Pearls had weight in the Dutch Republic. Fisheries in overseas territories yielded handsome tax revenue as well as harvests. They were weighed in commercial transactions and figuratively weighed the character of their wearers. But in the paintings of Vermeer and his contemporaries, they also have atomic weight. When X-rayed, these painted pearls appear solid white among ghostly images of faces. The gems they represent are so much more fragile than diamonds or emeralds, but in a painting, they are literally a lead weight. Vermeer's white pigments show up as areas of dense white on an X-ray because they are made of lead. With its high atomic weight, lead resists the passage of X-rays and is radiopaque.

Johannes Vermeer, *Girl with a Pearl Earring*, c. 1665, oil on canvas.

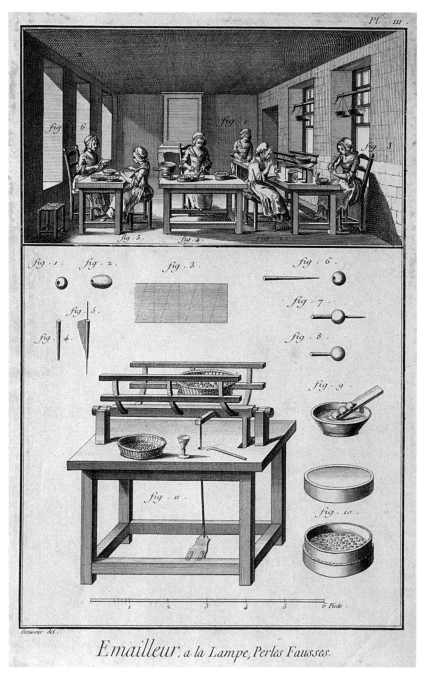

Louis-Jacques Goussier, *A Workshop for Artificial Pearls*,
French, 18th century, engraving by A. J. Defehrt.

In terms of technical art history, one of the most thoroughly an-
alysed paintings in the world is Vermeer's *Girl with a Pearl Earring* of
circa 1665. A recent investigation by an international team of scientists,
researchers, curators and conservators, with access to state-of-the-art
imaging technologies, has peeled back the layers of this painting to
determine just how Vermeer captured his famed pearly light, as well
as the pearls themselves.[38] We will never know what Vermeer titled
this work; for a while, it was known as *Girl with a Turban*.[39] But the
pearl eventually bested the drapery, and now our eyes are conditioned
to linger more on the voluminous gem suspended from her left ear
than on the cloth wrapped around her head.

Under magnification, the pearl's highlight is a thick daub of
lead white pigment on a drift of grey, like the Moon in front of a
cloud. For years, viewers saw a second highlight at the bottom of
the pearl. A 1994 restoration revealed that the highlight was the back
of a paint chip, possibly dislodged during restoration treatments in
1882.[40] Vermeer may have bought his pigments in his home city of
Delft from the apothecary Dirck de Cocq, who is known to have
stocked painters' staples such as lead white and linseed oil. But the
local supply of pigments, like pearls, owed as much to the Dutch
Republic's trading prowess. The rosiness of the girl's skin comes in
part from cochineal, ground from female insects which lived on
prickly pear cacti in Mexico. The black void on which she floats was
originally dark green, mixed from yellow and indigo, which could
have been shipped with cargoes of tobacco, sugar and spices from
the Caribbean. Her turban's extravagant ultramarine was extracted
from lapis lazuli, mined in the mountains of northeast Afghanistan.[41]
Lead isotope analysis, which can pin down a material's geographic
source, reveals that the lead ore in Vermeer's lead white came from
the Peak District in England.[42]

Until the nineteenth century, lead white was a commonly used
pigment for easel painting, valued for its clean opacity. The Dutch
had refined a way to process lead into pigment, known as the Dutch
stack process, in which sheets of lead were rolled up in clay pots of
vinegar. The pots were stacked in horse manure in a shed, and over

several weeks the chemical reaction propelled by heat and carbon dioxide produced snowy flakes of lead carbonate, which were ground into a powder. But lead white isn't just one pigment. The Dutch stack process produces chemicals of slightly different compositions, each with its unique optical properties. At least four different types of lead white pigments have been identified in *Girl with a Pearl Earring*.[43] From her flawless skin, to the whites of her eyes, to her laundered collar, to the highlight on her pearl, Vermeer chose different lead whites to capture different qualities of light.

Like any colour we see, white is relational; its effect shifts depending on the colours around it. The pearl gleams from the shadow of the girl's neck, which Vermeer under-painted in iron-rich, warm-toned earth pigments. Its cool sheen is enhanced by the blue turban above it and the mat silver on its base, reflecting her collar. Vermeer used a dab of white impasto as the finishing touch to the girl's blue-grey iris, twinning the animation of her eyes and the pearl.

The identity of the sitter remains a mystery, though her turban gives a clue. The work is a *tronie*, a type, a fantasy, in this case, a fantasy of an unspecified exotic East, the Orient, the place of pearls. Perhaps that explains the outsize earring, which turns out not to be an earring after all; it has no hook, except the one our eyes mistakenly supply.[44] It might not have been a pearl at all. It is so extravagantly large that if it existed, it was probably a manufactured bead. For centuries, chemists concocted artificial pearls from glass beads coated with mercury or a mixture of pearl powder and binder. Shortly before Vermeer painted *The Girl*, a French rosary-maker devised a substance he called *essence d'Orient*. More prosaically, it was made from fish scales suspended in oil. Mixed with adhesive, *essence d'Orient* could be blown into hollow glass beads to create large, lustrous artificial pearls.[45] Vermeer has already tricked us into seeing an earring that isn't there. His lead-white pearl is an illusion of a beautiful illusion.

Vermeer doesn't describe pearls' shape or colour but paints them as flickers of light. *Woman Holding a Balance* (c. 1664) was long known as *Girl Weighing Pearls*. But under the microscope, the balance was discovered to be empty. The hard, white highlights on the metal,

Johannes Vermeer, *A Young Woman Standing at a Virginal*, c. 1670–72, oil on canvas.

Muhammad Faqirullah Khan, *Women Playing on a Rock Slide*,
c. 1760, pigment, ink and gold on paper.

long thought to be pearls, are just reflections of light from a window. The strands of pearls on the table and draped over the box are painted differently. First, Vermeer laid down a soft grey band, then on top painted white highlights, each a slightly different size and shape, to suggest pearls' soft luminescence.[46] The strands are of subtly different colours; the pearls on the table are ivory, while the ones behind are silvery. In his later paintings, he pushed this technique even further. In *A Young Woman Standing at a Virginal* (*c.* 1670–72), he painted a thin, warm-grey band, blending with the sitter's skin tones. The pearls are discs of white impasto that fuse at the back of her neck in full sunlight. On her throat, in shadow, there's only the grey band, yet Vermeer conjures pearls.

ABRACADABRA IN AWADH

If Vermeer is the magician who makes us see things that don't exist, there are Mughal artists who pull rabbits from hats to bring pearls to life on the page.

By the mid-eighteenth century, the historical province of Awadh in northeastern India had arrived. Because of its rich alluvial soil, it had long been an important agricultural hub. The region became

Detail of Muhammad Faqirullah Khan, *Women Playing on a Rock Slide*, *c.* 1760, pigment, ink and gold on paper.

part of the Mughal Empire in the sixteenth century, and during the eighteenth, the Awadhi capital cities of Lucknow and Faizabad transformed into artistic powerhouses. In part, this was due to the erosion of the Mughal Empire outside Awadh. Delhi, the Mughal capital, was sacked in 1739 by the Iranian conqueror Nadir Shah, then suffered waves of invasion later that century. Intellectuals, poets, musicians and artists joined streams of migrants to the affluent and peaceful province of Awadh. At Faizabad and Lucknow, they helped create dazzling, multicultural communities of South Asians and Persians, Mughal émigrés, refugees, European expatriates, world travellers and traders.

Delhi's loss was Awadh's gain, though plenty of nostalgic exiled poets found much to disparage. Mir Taqi Mir bitterly reflected,

> Far better than Lucknow the ruins of Delhi:
> Would that I had died back there
> than let my madness lead me here![47]

Delhi-snob he may have been, but he managed to thrive for thirty years in Lucknow, contributing to its literary blossoming supported by Mughal patronage.

Other exiles reconciled themselves to their new home and even learned to revel in the many benefits of Awadh's thriving cities. The Delhi poet Mir Hasan found Faizabad a 'flower garden', with 'open bazaars and wide-open streets/ like lines on a white piece of paper'.[48] The bazaars of Lucknow and Faizabad offered sorrowful exiles plenty of consolation. There were flower stalls heaped with fragrant jasmine and bright marigolds; food stalls offering fruits, spices, sweets, cashews and kebabs; gem merchants selling emeralds, lapis lazuli and pearls; perfumiers, embroiderers and art dealers. There were coffee shops, brothels and women drenched in perfume and sandalwood.[49]

The artist Muhammad Faqirullah Khan was another displaced Delhi luminary. An imperially trained painter, he was able to secure patronage in the emerging artistic centres of Awadh. These were cities of palaces and gardens, courtiers and courtly love. The artist shows us princesses on terraces plucking blossoms while behind them stroll peacocks, cranes and deer. They wear robes of the finest *chikan* work, with shawls stitched in lavish gold *zardozi* embroidery. Pearls, rubies and emeralds glisten in their hair and dangle from their ears, and ropes of pearls circle their necks.

Historically, men and women at the Mughal court consumed vast quantities of pearls. They were such connoisseurs that the imperial treasury sorted pearls into sixteen classes.[50] The finest were reserved for emperors, who wore them wrapped around their torsos and wrists, on their fingers, and hanging in thick tassels from their belts. The artists and patrons who migrated to Awadh brought their love of pearls and their Mughal artistic heritage with them.[51] Awadhi painting, like its culture, was a rich infusion of diverse influences, including European conventions of perspective and space.

Faqirullah Khan's *Women Playing on a Rock Slide* would originally have been part of an album. The painting is framed by album page borders, with vine motifs picked out in gold. On the folio's reverse is Persian calligraphy. It is a self-contained world, alluding to music, poetry, gardens, refined leisure and the Awadhi court's famed opulence in textiles, dress and jewellery. Mughal miniatures are painted in opaque watercolours with organic and inorganic pigments such as indigo, ultramarine from lapis lazuli, vermilion (mercury sulphide), red lead and verdigris (copper chloride). Lead white is commonly used for depicting pearls, usually painted as white circles pin-pricked upwards to create dimension. On this album page, though, real seed pearls are used for every woman's jewellery, and possibly real rubies and emeralds as well.[52]

Typical of Mughal miniature painting, its surface is burnished. Mughal folios were made of Islamic paper constructed of hemp and flax and coated with starch. At various points during the painting process, the paper would be rubbed from the reverse with a smooth stone, burnishing the page so that its surface was as glossy as a pearl.

By including real pearls, the artist calls up a place where life imitates art. Nourished by memories of their beloved Delhi, their senses opened by a new, lush landscape and a culture of reinvention, Awadh's exiled artists set themselves to the task of remaking the world in art's image. As Mir Hasan wrote of a generous patron who underwrote this enterprise:

> Bravo! See the tastes and desires of lovers,
> Songs in an assembly, and the recollections of friends.
> It is one of the exemplars of Hindustan,
> It is a page from that wonderful album.[53]

PAINTING THE FLAMING PEARL

The mollusc-rich waters around the Ryukyu Islands, which arc gracefully between the southern tip of Japan and Taiwan, provided a bounty of mother-of-pearl, the iridescent nacre secreted on the inside of pearl-bearing molluscs' shells. Situated on maritime trading routes between Japan, China and Southeast Asia, the Ryukyus were infused by these regions' artistic traditions. The islands exported indigenous raw materials, one of the most valuable being tough, shiny shell from *Turbo cornutus*, a marine snail that flourished in these waters. Islanders used the shell as inlay for lacquerware, a traditional craft drawing on Chinese and Japanese prowess. It became so central to the regional economy that it was controlled by a Government Authority for Shell Polishing, which first appeared in official records in the early seventeenth century.[54] By the eighteenth century, the islands' artisans supplied lacquerware inlaid with indigenous mother-of-pearl to the Qing court in China, where sophisticated connoisseurs appreciated the conceit of representing a pearl with shimmering mother-of-pearl.

The use of mother-of-pearl as a decorative device stretches back millennia. As mother-of-pearl is far less rare than spherical pearls, it was, unsurprisingly, used abundantly across cultures. In Ur in ancient Mesopotamia, artisans combined it with lapis lazuli, copper, shale or limestone to create dazzling inlays for temple columns, musical instruments and domestic objects.[55] In pre-Columbian Mexico, Aztec artists contrasted its white sheen with turquoise and

Unknown artist, *Dish with Two Dragons Chasing a Flaming Pearl,* Ryukyu Islands, c. 1700–1800, black lacquer on wood core with mother-of-pearl inlay.

Unknown artist, box, Turkey, Ottoman, c. 1640, wood, tortoiseshell, mother-of-pearl, ivory and bone inlay.

malachite. Mother-of-pearl's potential as a glittering accent has been exploited from Song dynasty China to art nouveau Austria. Skilled cabinetmakers combined it with tortoiseshell, bone, silver, ivory and rare woods to transform the surfaces of furniture into marquetry of dazzling artistry. At the DIY end of the spectrum, Victorian women were encouraged to take up the art of papier mâché inlaid with mother-of-pearl to make small decorative goods such as book covers.[56]

In China, inlaying lacquer with mother-of-pearl was an ancient craft; luxurious inlaid red lacquer has been excavated from twelfth and thirteenth century BCE tombs of Shang dynasty royalty.[57] During the Tang dynasty, artisans contrasted glossy *T. cornutus* with aromatic sandalwood, amber-coloured tortoiseshell, jade, gold and silver. From the late Yuan period (fourteenth century), the focus switched to perfecting figurative illusions. Instead of the thick plates of mother-of-pearl typical of Tang work, thousands of delicate, wafer-thin fragments from the snail *Haliotis tuberculata* were combined to represent blossoming plum trees in moonlight, palace gardens, Daoist paradises and Buddhist legends. A seventeenth-century lacquer master explained, 'The greater the care that is taken with the details of inlays in creating a similarity to painting, the better.'[58]

In Japan, the application of mother-of-pearl (*raden*) found its perfect foil in shiny black lacquer, a natural plastic made from the sap of the *Toxicodendron vernicifluum* tree. Artisans devised ingenious techniques for making the shell imitate flower petals, water, even fog. They sprinkled it through templates, layered it in thin, delicate fragments, applied it in thick plates, arranged it into geometric, coloured mosaics, and manipulated its colours by layering it over metal foil or pigments.[59]

The Ryukyus absorbed artistic influences from trading partners just as the islands' mountains drew down the plentiful rains. Large platters such as the one overleaf were presented as tribute from the islands to the Chinese court in Beijing. A pair of Chinese dragons – symbols of energy and regeneration – emerges in shimmering mother-of-pearl, scales minutely incised, claws shimmering like rainbows. Sometimes five different types of the shells so plentiful on these islands were combined into images that flickered between green, pink and turquoise – creatures of mist and water, generated from matter born in water. The dragons toss between them a flaming pearl, a symbol of wisdom constantly replenished like nature's perfect gem, endlessly renewed in the oceans lapping the Ryukyus.

THE PILGRIM'S PROCESS

At first glance, the American photographer Catherine Opie is the photo negative of Elizabeth Taylor – Taylor the diva in diamonds, Opie the pragmatist in T-shirt and tats. Or less so, in her famous self-portrait in a latex bondage mask with the word *Pervert* carved above her bare breasts. The women seem as mismatched as Chanel and Vans footwear. And yet, without ever meeting, they colluded on a portrait as poignant and iconic as William Eggleston's images of Graceland six years after Elvis's death.

In early 2011 Opie began photographing Taylor's possessions in her Bel Air home at 700 Nimes Road, Los Angeles. Over six months, she created nearly 3,000 images of the star's cherished belongings, from Sugar Ray Robinson's autographed boxing gloves to a Warhol sketch (signed 'to Elizabeth, a big kiss'), to shelves of intricately tooled cowboy boots, racks of couture, rows of photographs of favourite dogs and children, a clutter of cosmetics and perfumes.[60] It was Opie's 'intimate portrait of Elizabeth Taylor through her personal space'.[61] Halfway through the project, Taylor died, and the portfolio transformed into a memorial.

The project, initiated by Opie, an admirer of Taylor's AIDS activism, was at times discomfiting for the photographer whose breakout work documented the lesbian bondage community in 1980s San Francisco. She recalls that, among the handbags and perfume bottles, 'my identity as a butch woman was challenged, and sometimes I would put my scruffy tennis shoes next to a Chanel pump and shake my head in awe.'[62]

The publication that emerged from this project, *700 Nimes Road*, distils these shots to a collection of perfectly composed images

Catherine Opie, *700 Nimes Road, La Peregrina (Elizabeth Taylor)*, 2012, pigment print.

telescoping from *Vogue*-esque wide-angled interiors to intimate close-ups of a fur cuff on a brocade sleeve. 'I do like to stare,' admits Opie, and through her lens, these tightly cropped objects have the look of exotic animals caught in her crosshairs.[63]

Some of the most arresting images capture Taylor's fabled jewellery collection, packed up by Christie's for auction. There is the heart-shaped diamond Richard Burton gave to Taylor on her fortieth birthday, which was previously owned by another enamoured husband, Shah Jahan, the Mughal emperor who built the Taj Mahal.[64] And the flawless Krupp Diamond, named after the family of German industrialists. There are rubies, sapphires and diamonds from Bulgari, and the Van Cleef & Arpels diamond daisy suite Taylor wore when she received the Jean Hersholt Humanitarian Award for her AIDS activism.

And there is La Peregrina, possibly the world's most storied pearl. Unlike many of the other photographs, where Opie's depth of field captures gems in all their crisp brilliance, the image of La Peregrina is a milky blur, its Cartier setting a scattered froth of bokeh. It's a photograph of the idea of the jewel, just as the posthumous portrait of Taylor turned out to be a portrait of her aura.[65]

La Peregrina (The Pilgrim) has slipped in and out of history many times; it evaporated from public view in 2011 after being sold at auction for an eye-watering $11,842,500.[66] To everyone except its anonymous buyer, it exists only as an aura, captured in the written record, in paintings, in memory, in photographs, but most of all in the imagination. Even the details of its discovery are disputed. It was found at the start of the sixteenth century by a black slave who was rewarded with his freedom. It was found in 1560. Or the 1570s. It came from the Gulf of Panama. It came from the coast of Venezuela in 1574.[67] Its beginnings are as hazy as haar. Its size (200 grains, or 10 grams) suggests that it was probably retrieved from Panamanian waters, as the local oyster *Pinctada mazatlanica* produces larger pearls than Venezuelan species.[68] In the early sixteenth century, Spanish colonizers discovered gold and pearls along the Panama isthmus. They were given, or stole, stashes of pearls and named one of the islands off the southern coast 'Isla de las Perlas'.[69]

How and when the pearl made its way to Spain is another mystery. Some reports claimed it became part of the Spanish Royal Treasury in 1513, after Vasco Núñez de Balboa, the first European to cross Panama and reach the Pacific, brought it back to King Ferdinand V.[70] In another account, the explorer Don Diego de Temes brought it back for Philip II.[71]

The gossipy diplomat Louis de Rouvroy, duc de Saint-Simon, had the chance to see La Peregrina in person in 1721 and said it reminded him of the tiny, candied muscat pears sold by Parisian confectioners.[72] People who saw La Peregrina imagined putting it in their mouths. This is the icing-sugar 'mouth-feel' captured by Opie. Her exaggerated soft focus is achingly romantic. Striped of identifying features, it embodies every pedigreed pearl handed down by kings and their favourites, gracing the painted necks of queens, stuffed into the pockets of desperate aristocrats dodging the guillotine or the Bolsheviks. Like our best aspirations, it shimmers just out of reach.

It wasn't only people who found La Peregrina mouth-watering. After tempting the long line of its royal owners in Spain and France, it ended up on the auction block in London in 1969 and was bought by Richard Burton as a Valentine's Day present for his wife, Elizabeth Taylor. During a visit to Caesar's Palace in Las Vegas, she reached to stroke the pearl and realized with horror that it was missing. After a panicked scrabble through the carpet, she noticed her Pekingese puppy crunching an object, prised open its jaws and retrieved 'the most perfect pearl in the world. It was – thank God – not scratched.'[73] The episode didn't deter Taylor's famous love of lush carpeting, captured in Opie's portrait.

After her death, just before Christie's took Taylor's possessions away for auction, Opie and her team put the jewels out in the sun 'to mark a moment of silence'.[74] The images from that morning are like objects brought back from a pilgrimage. Among them glows La Peregrina, tempting our belief that we know this legendary pearl, just as we 'know' its celebrity owner. But, in both offering and withholding the pearl's perfection, the photograph points to the animal and human lives we fall short of grasping. It reminds us that people, like pear-shaped pearls, are not for consuming.

References

PREFACE Sarah Siddons's Necklace

1 For a chronology of Siddons's life, see Robyn Asleson, ed., *A Passion for Performance: Sarah Siddons and Her Portraitists* (Los Angeles, CA, 1999), pp. xii–xvii.

2 Quoted in Thomas Campbell, *Life of Mrs Siddons* (London, 1834), vol. I, pp. 243–4.

3 Shelley Bennett and Mark Leonard, with technical studies by Narayan Khandekar, '"A Sublime and Masterly Performance": The Making of Sir Joshua Reynolds's *Sarah Siddons as the Tragic Muse*', in *A Passion for Performance*, ed. Asleson, p. 124.

4 Bennett et al., 'A Sublime and Masterly Performance', p. 131.

5 Quoted in Campbell, *Life of Mrs Siddons*, p. 242.

6 Marcia Pointon notes that jewels were regularly borrowed, in *Brilliant Effects: A Cultural History of Gem Stones and Jewellery* (London, 2009), p. 24.

7 Asleson, ed., *A Passion for Performance*, p. xvi.

8 William Hazlitt, 'Mrs Siddons', *The Examiner*, 16 June 1816, in *The Complete Works of William Hazlitt*, ed. P. P. Howe (London, 1930–34), vol. V, p. 312.

9 The Royal Collection owns one of the numerous copies of this portrait, RCIN 402413.

10 See Robyn Asleson, ed., *Notorious Muse: The Actress in British Art and Culture, 1776–1812* (New Haven, CT, 2003).

11 Jonathan Bate, 'Shakespeare and the Rival Muses: Siddons versus Jordan', in *Notorious Muse: The Actress in British Art and Culture, 1776–1812* (New Haven, CT, 2003), ed. Robyn Asleson, pp. 81–103 (p. 95).

12 The seventeenth-century writer Solomon Richards on the pearls of Wexford, Ireland, quoted in George Frederick Kunz and Charles Hugh Stevenson, *The Book of the Pearl: The History, Art, Science, and Industry of the Queen of Gems* (New York, 1908), p. 162.

ONE The Oyster's Autobiography

1 Legends about pearls' formation are discussed by George Frederick Kunz and Charles Hugh Stevenson, *The Book of the Pearl: The History, Art, Science, and Industry of the Queen of Gems* (New York, 1908), pp. 3–5; R. A. Donkin, *Beyond Price: Pearls and Pearl Fishing, Origins to the Age of Discoveries* (Philadelphia, PA, 1998), pp. 1–22; Elisabeth Strack, *Pearls* (Stuttgart, 2006), pp. 18–20.

2 Quoted by Eugene Walter, 'Federico Fellini, Wizard of Film', *The Atlantic*, CCXVI/6 (December 1965), p. 67.

3 Lewis Carroll, *Alice's Adventures in Wonderland* [1865] (London, 2007), p. 39.

4 Jason Daley, 'See the World's Oldest Pearl, Soon to Go on View for the First Time', *Smithsonian Magazine*, www.smithsonianmag.com, 23 October 2019.

5 In 2011, for example, a marine pearl was recovered from 2,000-year-old layers in a shell midden at Brremangurey, on the Kimberley coast, Australia. See Katherine Szabo et al., 'The Brremangurey Pearl: A 2,000 Year Old Archaeological Find from the Coastal Kimberley, Western Australia', *Australian Archaeology*, LXXX (June 2015), pp. 112–15.

6 Francis Ponge, 'L'huître', published in French in *Le Parti Pris des Choses*, 1942, trans. Yasmine Seale, www.theislandreview.com, accessed 13 June 2021.

7 For bivalve anatomy, see Shigeru Akamatsu, *Pearl Book* (Tokyo, 2015), pp. 59–61.

8 Very occasionally, non-nacreous pearls have been discovered in edible oysters. See, for example, Britni LeCroy and Artitaya Homkrajae, '15.53 Ct. Pearl Discovered in Edible Oyster from the Ostreidae Family', *Gems and Gemology*, LVI/3 (Fall 2020), pp. 420–22.

9 For an in-depth discussion of species of pearl-bearing mollusc, see Neil Landman et al., *Pearls: A Natural History* (New York, 2001), pp. 23–39; Strack, *Pearls*, pp. 41–112.

10 *Pinctada radiata* is so closely related to *P. fucata* (the Japanese Akoya oyster) that they have, by some zoologists, been considered the same species.

11 The International Union for the Conservation of Nature (IUCN) classified this pearl mussel as Critically Endangered in Europe. In the United Kingdom it is protected under the UK Wildlife and Countryside Act 1981, and under the UK Biodiversity Action Plan is listed as a 'Priority Species' requiring an action plan dedicated to its survival.

12 Annual production figures for the decade 2005–14, from Changbo Zhu, Paul C. Southgate and Ting Li, 'Production of Pearls', in *Goods and Services of Marine Bivalves*, ed. Aad C. Smaal et al. (Cham, 2019), p. 73.

13 Strack, *Pearls*, p. 233.

14 Ibid., p. 242.

15 The Portuguese explorer Pedro Teixeira (1586–1605) and the Flemish naturalist Anselmus de Boodt (1550–1632) wrote on the similarity between a mollusc's shell and its pearl in 1608 and 1609, respectively.

16 The composition of pearls can differ, with some containing a higher percentage of conchiolin. See Landman et al., *Pearls*, p. 42; Strack, *Pearls*, p. 285.

17 Plato, *Phaedrus*, 250c, from *Plato in Twelve Volumes*, vol. IX, trans. Harold N. Fowler (London, 1925), www.perseus.tufts.edu, accessed 26 December 2020.

18 Kunz and Stevenson discuss the prevalence of this theory, *Book of the Pearl*, pp. 37–9.

19 Kiyohito Nagai, 'A History of the Cultured Pearl Industry', *Zoological Science*, XXX/10 (2013), pp. 783–93 (p. 786).

20 See Akamatsu, *Pearl Book*, pp. 57–62 for a technical discussion of the mechanism of pearl growth; Strack, *Pearls*, pp. 116–17; Landman et al., *Pearls*, pp. 27–8.

21 Strack, *Pearls*, p. 38.

22 Christie's Auction 2623, lot 12, 12 December 2011; Sotheby's Auction GE1809, lot 100, 14 November 2018.

23 Quran 22.23.

24 Matthew 13:45–6, *Zondervan NIV Study Bible* (Grand Rapids, MI, 2011), p. 1615.

25 Carol Ann Duffy, 'Warming Her Pearls', in *Collected Poems* (London, 2015), p. 120.

26 Strack, *Pearls*, p. 115.

27 Stephen G. Bloom, *Tears of Mermaids: The Secret Story of Pearls* (New York, 2009), p. 4.

28 Gemological Institute of America, 'The 7 Pearl Value Factors', www.gia.edu, accessed 10 December 2020.

29 Al-Bīrūnī, *The Book Most Comprehensive in Knowledge on Precious Stones*, trans. Fritz Krenkow, ed. Hakjim Mohammad Said (Islamabad, 1989), p. 104.

30 Donkin, *Beyond Price*, p. 179.

31 John Steinbeck, *Cannery Row* [1945] (New York, 1992), p. 78.

32 Al-Bīrūnī, *The Book Most Comprehensive*, pp. 108, 103.

33 Landman et al., *Pearls*, pp. 41–6; Strack, *Pearls*, pp. 119–23; Akamatsu, *Pearl Book*, pp. 57–8.

34 Richard Wise, *Secrets of the Gem Trade: The Connoisseur's Guide to Precious Gemstones* (Lenox, MA, 2004), p. 70.

35 Landman et al., *Pearls*, p. 48; Strack, *Pearls,* pp. 289–92.

36 Al-Bīrūnī, *The Book Most Comprehensive*, p. 94.

37 Kalpana S. Katti and Dinesh R. Katti, 'Why is Nacre so Tough and Strong?', *Materials Science and Engineering*, XXVI/8 (2006), pp. 1317–24.

38 Biophysicist and geobiologist Pupa Gilbert investigates the poorly understood formation of nacre at the molecular level and theorizes its structure derives from environmental conditions during the period of nacre deposition. See, for example, I. C. Olson, R. Kozdon, J. W. Valley and Pupa Gilbert, 'Mollusk Shell Nacre Ultrastructure Correlates with Environmental Temperature and Pressure', *Journal of the American Chemical Society*, CXXXIV/17 (2012), pp. 7351–8.

39 Zhu et al., 'Production of Pearls', p. 87.

40 Helen Scales, *Spirals in Time: The Secret Life and Curious Afterlife of Seashells* (London, 2015), pp. 247–9.

41 Because the commercial exploitation of pearl oysters is of comparatively higher value to more countries than the exploitation of pearl-bearing

freshwater mussels, more research has been conducted on the effects
of salinity and temperature on these species.

42 For a discussion about the building of shell in warm and cold waters,
    see Cynthia Barnett, *The Sound of the Sea: Seashells and the Fate of the
    Oceans* (New York, 2021), pp. 48–9.

43 Gunawan Muhammada et al., 'Nacre Growth and Thickness of Akoya Pearls
    from Japanese and Hybrid *Pinctada fucata* in Response to the Aquaculture
    Temperature Condition in Ago Bay, Japan', *Aquaculture*, CDLXXVII (2017),
    pp. 35–42.

44 Pupa Gilbert et al., 'Nacre Tablet Thickness Records Formation Temperature
    in Modern and Fossil Shells', *Earth and Planetary Science Letters*, CDLX
    (2017), pp. 281–92.

45 Quran 55.19–23. See Robert A. Carter, *Sea of Pearls: Seven Thousand Years
    of the Industry that Shaped the Gulf* (London, 2012), pp. 32–3.

46 Andrew M. Gardner, *City of Strangers: Gulf Migration and the Indian
    Community in Bahrain* (Ithaca, NY, 2010), p. 35.

47 John S. Lucas, 'Environmental Influences', in *The Pearl Oyster*,
    ed. Paul C. Southgate and John S. Lucas (Amsterdam, 2008),
    pp. 187–229.

48 Laura Payton and Damien Tran, 'Moonlight Cycles Synchronize
    Oyster Behaviour', *Royal Society Biology Letters*, 15:20180299 (2019),
    doi.org/10.1098/rsbl.2018.0299.

49 Lucas, 'Environmental Influences', p. 216.

50 Al-Bīrūnī, *The Book Most Comprehensive*, pp. 102–4. Carter compiles
    all the terms used by al-Bīrūnī in *Sea of Pearls*, pp. 290–91.

51 Kunz and Stevenson, *Book of the Pearl*, p. 321.

52 Ibid., p. 353.

53 Akamatsu, *Pearl Book*, p. 141, notes this term.

54 Landman et al., *Pearls*, p. 58.

55 At retail level, pearls are commonly assigned ratings up to AAAA, in part
    determined by their roundness.

56 The Natural History Museum, London, has an oyster specimen donated
    in 1886 with a pearlfish trapped in nacre. See Emily Osterloff, 'The Fish
    That's also a Pearl', www.nhm.ac.uk, accessed 30 December 2020.

57 Kunz and Stevenson, *Book of the Pearl*, p. 352.

58 Molly Warsh, *American Baroque: Pearls and the Nature of Empire, 1492–1700*
    (Chapel Hill, NC, 2018), p. 9.

59 The physiological origin of these grooves is still unclear, though many
    scientists hypothesize that they result from uneven rotation within the
    pearl sac.

60 Landman et al., *Pearls*, p. 56.

61 Kunz and Stevenson, *Book of the Pearl*, p. 352.

62 Strack, *Pearls*, pp. 294–5.

63 Kunz and Stevenson, *Book of the Pearl*, p. 352.

64 For an entertaining account of this pearl's misadventures, see Michael
    LaPointe, 'Chasing the Pearl of Lao Tzu', *The Atlantic* (June 2018),
    www.theatlantic.com, accessed 4 January 2021.

65 Alfonsina Storni, 'You Want Me White', trans. Anabella S. Paiz, *Prairie Schooner*, LXXI/3 (Fall 1997), p. 171.

66 Al-Bīrūnī, *The Book Most Comprehensive*, pp. 97–8.

67 Camillo Leonardi, *The Mirror of Stones* [1502] (London, 1750), pp. 200–201.

68 Patrick Baty, *The Anatomy of Colour: The Story of Heritage Paints and Pigments* (London, 2017), pp. 86–8.

69 George Field, *Chromatography: A Treatise on Colours and Pigments, and of their Powers in Painting* [1835] (London, 1885), p. 89; Arthur Seymour Jennings, *Paint and Colour Mixing: A Practical Handbook for Painters, Decorators, Artists* (London, 1906), p. 86.

70 'The Munsell Book of Color 1929: Traditional Color Names', www.munsell.com, accessed 12 January 2021.

71 Ian Paterson, *A Dictionary of Colour: A Lexicon of the Language of Colour* (London, 2004), p. 293.

72 Pantone Pearl 12-1304.

73 This is an area that is currently widely researched in relation to cultured pearls, in part because of the economic importance of pearl colour. See, for instance Chin-Long Ky et al., 'Impact of Spat Shell Colour Selection in Hatchery-Produced *Pinctada margaritifera* on Cultured Pearl Colour', *Aquaculture Reports*, IX (2018), pp. 62–7; Ziman Wang et al., 'How Cultured Pearls Acquire Their Colour', *Aquaculture Research*, 51 (2020), pp. 1–10.

74 I am grateful to Devchand Chodhry, Chairman of Orient Pearl (Bangkok) Ltd, for discussion on the factors influencing the colour of cultured pearls, interview 7 January 2021.

75 *New York Times*, 12 November 1936, p. 3.

76 Lucas, 'Environmental Influences', p. 221.

77 Pliny the Elder, *The Natural History*, 9.54, ed. John Bostock and H. T. Riley, in the Perseus Digital Library, www.perseus.tufts.edu, accessed 7 January 2021.

78 Katsuhiko T. Wada and Dean R. Jerry, 'Population Genetics and Stock Improvement', in *The Pearl Oyster*, ed. Paul C. Southgate and John S. Lucas (Oxford, 2008), pp. 452–4.

79 Strack notes the industry-wide use of treatments such as bleaching, dyeing and polishing, *Pearls*, pp. 649–64.

80 Al-Bīrūnī discusses whiteness and the desirability of a yellow tinge at length, *The Book Most Comprehensive*, pp. 95–8.

81 Lewis Pelly, 'Remarks on the Pearl Oyster Beds in the Persian Gulf' [1866], in *Records of the Persian Gulf Pearl Fisheries, 1857–1962*, vol. I: *1857–1914*, ed. Anita L. P. Burdett (Cambridge, 1995), p. 10.

82 Hubert Bari, David Lam and Susan Hendrickson, *The Pink Pearl: A Natural Treasure of the Caribbean* (Milan, 2007), p. 69.

83 Bram Hertz, *A Catalogue of the Collection of Pearls and Precious Stones Formed by Henry Philip Hope, Esq.* (London, 1839), p. 8.

84 Emmanuel Fritsch and Elise B. Misiorowski, 'The History and Gemology of Queen Conch "Pearls"', *Gems and Gemology*, XXIII/4 (Winter 1987), pp. 209–10.

85 See Kunz and Stevenson, *Book of the Pearl*, p. 60.
86 Strack, *Pearls*, p. 289.
87 George Frederick Kunz, 'The Crown Jewels of Russia', *Art and Life*, x/2 (June 1919), pp. 288–92.
88 For an account of the black pearl industry in French Polynesia, see Colin Newbury, *Tahiti Nui: Change and Survival in French Polynesia, 1767–1945* (Honolulu, HI, 1980).
89 Strack, *Pearls*, p. 528. Bloom records extensive conversations with Assael in *Tears of Mermaids*.
90 Landman et al., *Pearls,* p. 191.
91 For instance, *A Hand-Book for Travellers in Southern Germany* (London, 1844), p. 87.
92 Russell Shor, 'From Single Source to Global Free Market: The Transformation of the Cultured Pearl Industry', *Gems and Gemology*, XLIII/3 (Fall 2007), p. 214.
93 Laurent E. Cartier and Saleem Ali, 'Pearl Framing as a Sustainable Development Path', *Solutions for a Sustainable and Desirable Future*, III/4 (August 2012), pp. 30–34; Shor, 'From Single Source to Global Free Market', p. 214.
94 Strack, *Pearls*, p. 536.
95 Wayne O'Connor and Scott P. Gifford, 'Environmental Impacts of Pearl Farming', in *The Pearl Oyster*, ed. Paul C. Southgate and John S. Lucas (Oxford, 2008), pp. 497–525.
96 Laurent E. Cartier and Kent E. Carpenter, 'The Influence of Pearl Oyster Farming on Reef Fish Abundance and Diversity in Ahe, French Polynesia', *Marine Pollution Bulletin*, LXXVIII/1–2 (2014), pp. 43–50; Julie Nash et al., 'The Sustainable Luxury Contradiction: Evidence from a Consumer Study of Marine-Cultured Pearl Jewellery', *Journal of Corporate Citizenship*, 63 (2016), pp. 73–95.
97 See the data and case studies presented by the research project Sustainable Pearls, based at the University of Vermont in collaboration with partners in Switzerland, the United States, Japan and the Pacific region. www.sustainablepearls.org.
98 See the case study on Kamoka Pearls, profiled by Sustainable Pearls, www.sustainablepearls.org, accessed 26 February 2021.

TWO Harvest

1 The following description of the 1932 riot was compiled from contemporary British government records relating to economic activity in the Gulf, published as *Records of the Persian Gulf Pearl Fisheries, 1857–1962*, ed. Anita L. P. Burdett (Cambridge, 1995), vol. III, pp. 53–98.
2 Britain's Political Agent in Bahrain, Captain Charles Geoffrey Prior, letter to the Political Resident in the Persian Gulf, 30 May 1932, 'File 35/3 The Divers' Riot of May 1932' [8v] (16/104), British Library: India Office Records and Private Papers, IOR/R/15/2/848, in Qatar Digital Library, www.qdl.qa, accessed 11 November 2021.

3   Robert A. Carter, *Sea of Pearls: Seven Thousand Years of the Industry that Shaped the Gulf* (London, 2012), p. 169.
4   Ibid., p. 3.
5   Paolo Biagi et al., 'Qurum: A Case Study of Coastal Archaeology in Northern Oman', *World Archaeology*, XVI/1 (June 1984), pp. 43–61.
6   See R. A. Donkin, *Beyond Price: Pearls and Pearl Fishing, Origins to the Age of Discoveries* (Philadelphia, PA, 1998), p. 47 for a discussion on Bahrain's role as an entrepôt.
7   *The Epic of Gilgamesh*, trans. and ed. Benjamin R. Foster (New York, 2001), p. 94.
8   Numerous texts have argued this passage is an early reference to pearl diving. See Donkin, *Beyond Price*, p. 48.
9   Donkin, *Beyond Price*, pp. 51–2.
10 Pliny the Elder, *The Natural History*, 37.6, ed. John Bostock and H. T. Riley, in the Perseus Digital Library, www.perseus.tufts.edu, accessed 22 March 2021.
11 See Burdett, *Records of the Persian Gulf Pearl Fisheries*, vol. III, pp. 53–98 for derogatory terms used of divers during this period.
12 Donkin, *Beyond Price*, p. 124. Over time, there was some fluidity in the length of the diving season, though the water in winter and early spring was too cold for sustained diving.
13 See the maps published as vol. IV, in Burdett, *Records of the Persian Gulf Pearl Fisheries*.
14 The British civil servant John Gordon Lorimer compiled a detailed account of the Persian Gulf pearl fisheries in the early twentieth century, included in his compendious *Gazetteer of the Persian Gulf, Oman and Central Arabia*, published by the British government in Calcutta in 1908 and 1915.
15 Neil Landman et al., *Pearls: A Natural History* (New York, 2001), p. 132.
16 Robert Carter, 'Pearl Fishing, Migration, and Globalization in the Persian Gulf, Eighteenth to Twentieth Centuries', in *Pearls, People and Power: Pearling and Indian Ocean Worlds*, ed. Pedro Machado et al. (Athens, OH, 2019), pp. 232–62 (p. 236).
17 This is an estimate by a senior British naval officer in the Gulf, quoted by Matthew S. Hopper, 'Enslaved Africans and the Globalization of Arabian Gulf Pearling', in *Pearls, People, and Power: Pearling and Indian Ocean Worlds*, ed. Pedro Machado et al. (Athens, OH, 2020), p. 267.
18 Carter in *Sea of Pearls* observed that the basic routine and equipment was unchanged over centuries, p. 31. Lorimer provided detailed eye-witness accounts in his *Gazetteer*.
19 Alan Villiers, *Sons of Sinbad* (New York, 1940), p. 378.
20 Lorimer, *Gazetteer*, notes this practice from the late nineteenth century.
21 Carter, *Sea of Pearls*, p. 14.
22 João Ribeiro, *The Historic Tragedy of the Island of Ceilão*, trans. P. E. Pieris (New Delhi, 1999), p. 72.
23 George Frederick Kunz and Charles Hugh Stevenson, *The Book of the Pearl: The History, Art, Science, and Industry of the Queen of Gems* (New York, 1908), p. 91.

24 Carter, *Sea of Pearls*, p. 63, notes that magicians were employed up to
   the nineteenth century in the Gulf of Mannar.

25 Ibn Battutah, *Travels in Asia and Africa, 1325–1454*, trans. H.A.R. Gibb
   (London, 1963), p. 122.

26 *Gilgamesh*, p. 95.

27 Ibid.

28 Villiers, *Sons of Sinbad*, p. 375.

29 A collection of these petitions dating from 1928 to 1934 is published in
   Burdett, *Records of the Persian Gulf Pearl Fisheries*, vol. III, pp. 99–209.

30 Ribeiro, *Historic Tragedy*, p. 73.

31 Quoted in Nasser Al-Taee, '"Enough, Enough, Oh Ocean": Music of the Pearl
   Divers in the Arabian Gulf', *Middle East Studies Association Bulletin*, XXXIX/1
   (June 2005), pp. 19–30 (p. 26).

32 For example, the *Persian Gulf Pilot*, a series of navigational guides issued
   by the Admiralty, London, lists the locations of these structures.

33 For a discussion on these structures and their relation to Gulf economies,
   see Ronald Hawker et al., 'Wind-Towers and Pearl Fishing: Architectural
   Signals in the Late Nineteenth and Early Twentieth Century Arabian Gulf',
   Antiquity, LXXIX/305 (2005), pp. 625–35. In recent years there has been
   renewed architectural interest in these structures for their potential to reduce
   energy costs.

34 William Clarence-Smith, 'The Pearl Commodity Chain, Early Nineteenth
   Century to the End of the Second World War', in *Pearls, People, and Power*,
   ed. Machado et al., pp. 31–54 (p. 36).

35 Ki Hackney and Diana Edkins, *People and Pearls: The Magic Endures*
   (New York, 2000), p. 68. During these years, a reputed royal provenance
   (even if inaccurate) substantially inflated prices.

36 Carter, 'Pearl Fishing, Migration, and Globalization', p. 234. For conversion
   to today's money, see Bank of England's 'Inflation Calculator',
   www.bankofengland.co.uk, accessed 27 March 2021.

37 Ribeiro, *Historic Tragedy*, pp. 250–56.

38 Carter, *Sea of Pearls*, pp. 236–7.

39 Quoted by Hopper, 'Enslaved Africans', p. 264.

40 Lorimer, 'The Pearl and Mother-of-Pearl Fisheries of the Persian Gulf',
   Appendix C, *Gazetteer*, published in Burdett, *Records of the Persian Gulf
   Pearl Fisheries*, vol. I, p. 304.

41 The Residency Agent, Shargah, letter to the Political Resident, Persian Gulf,
   15 July 1924, in Burdett, *Records of the Persian Gulf Pearl Fisheries*, vol. II,
   p. 467.

42 C. D. Belgrave, 'The Pearl Industry of Bahrain', 19 December 1928, in Burdett,
   Records of the Persian Gulf Pearl Fisheries, p. 545.

43 Victoria Penziner Hightower, 'Pearls and the Southern Persian/Arabian
   Gulf: A Lesson in Sustainability', *Environmental History*, XVIII/1 (January
   2013), pp. 44–59 (p. 53).

44 Villiers, *Sons of Sinbad*, p. 372.

45 Andrew M. Gardner, *City of Strangers: Gulf Migration and the Indian
   Community in Bahrain* (Ithaca, NY, 2010), p. 40.

46 In Kuwait, there is a modest revival of pearl-diving traditions, culminating in a festival each summer, though this falls under the category of heritage and tourism rather than industry.

47 Third clause of the first *capitulación* with Columbus, quoted in Molly Warsh, *American Baroque: Pearls and the Nature of Empire, 1492–1700* (Chapel Hill, NC, 2018), p. 20.

48 For Columbus's expectations of China, see Zhang Zhishan, 'Columbus and China', *Monumenta Serica*, XLI/1 (1993), pp. 177–87.

49 Donkin, *Beyond Price*, p. 313.

50 Ibid.; Warsh, *American Baroque*, p. 31.

51 The 'Costa de Perlas' appears on Juan de la Cosa's seminal map of America in 1500. See Warsh, *American Baroque*, p. 33.

52 Aldemaro Romero, 'Death and Taxes: The Case of the Depletion of Pearl Oyster Beds in Sixteenth-Century Venezuela', *Conservation Biology*, XVII/4 (August 2003), p. 1013–23 (p. 1016)..

53 For a discussion on the meanings the indigenous peoples of the Caribbean gave to pearls, see Nicholas J. Saunders, 'Biographies of Brilliance: Pearls, Transformations of Matter and Being, c. AD 1492', *World Archaeology*, XXXI/2 (1999), pp. 243–57.

54 Warsh, *American Baroque*, p. 35.

55 Warsh discusses the extent to which official mechanisms for taxing pearls failed, ibid., pp. 52–3.

56 Romero, 'Death and Taxes', p. 1016.

57 Ibid., p. 1017.

58 Bartolomé de Las Casas, *A Short Account of the Destruction of the Indies* [1542], ed. and trans. Nigel Griffin, with an introduction by Anthony Pagden (London, 2004), p. 3.

59 Las Casas, *Short Account*, p. 93.

60 Ibid., p. 94.

61 Warsh, *American Baroque*, p. 46.

62 Kevin Dawson, *Undercurrents of Power: Aquatic Culture in the African Diaspora* (Philadelphia, PA, 2018), p. 9.

63 Ibid., p. 64.

64 Romero, 'Death and Taxes', p. 1017; Michael Perri, '"Ruined and Lost": Spanish Destruction of the Pearl Coast in the Early Sixteenth Century', *Environment and History*, XV/2 (May 2009), pp. 129–61 (p. 136).

65 There have been failed initiatives to restore the site, as recorded in Venezuelan director Jorge Thielen Armand's 2015 short documentary *Flor de la Mar*, which follows the valiant attempts of local fishermen to protect the ruins.

66 Dawson, *Undercurrents of Power*, p. 70.

67 Dogs greeted Stephen G. Bloom on his trip to Cubagua as research for his book *Tears of Mermaids: The Secret Story of Pearls* (New York, 2009).

68 Romero, 'Death and Taxes', p. 1018.

69 Ibid.

70 Ibid., p. 1019.

71 For detailed calculations based on tax figures and contemporary records, see ibid., pp. 1018–20.

72 Landman et al., *Pearls*, p. 10, referencing fashion historians.

73 Garcilaso de la Vega, writing in the 1560s, quoted by Warsh, *American Baroque*, p. 81.

74 Priscilla Muller, *Jewels in Spain, 1500–1800* (Madrid, 2012), p. 56.

75 Lapidaries had long prescribed pearls for numerous ailments. Muller references the Valencian poet Jaime Roig, who mentions pearls crushed in *limonada*, ibid., p. 32.

76 Ibid., p. 71.

77 G. A. Bergenroth, ed., *Supplement to Volume I and Volume II, Letters, Despatches, and State Papers Relating to the Negotiations between England and Spain* (London, 1862), p. xlvii. www.british-history.ac.uk, accessed 9 April 2021.

78 Joanna of Castile is known to history as Juana La Loca. Certainly traumatized, her reputed insanity has been called into dispute. See Greg Wilkinson, 'Juana la Loca/"Joanna the Mad" (1479–1555): Queen of Castile and of Aragon – and Necrophiliac?', *British Journal of Psychiatry*, CCXVII/2 (2020), p. 449.

79 Suetonius, 'The Life of Julius Caesar', chapter 47, from *The Lives of the Twelve Caesars*, trans. J. C. Rolfe (Cambridge, MA, 1914), www.penelope.uchicago.edu, accessed 10 April 2021.

80 See Lindsay Shen, 'Mussel Memory: Reading History in the Layers of a Pearl', *Orion: People and Nature*, XL/2 (Summer 2021), pp. 24–31.

81 Gold finger ring containing a lock of Prince Charles Edward Stuart's hair given by him to Alexander Stuart of Invernayle, with seed pearls once forming the initials 'C. R.', National Museums Scotland, H.NJ 84.

82 See *A Collection of Inventories and Other Records of the Royal Wardrobe and Jewelhouse; and of the Artillery and Munitions in Some of the Royal Castles, 1488–1606*, ed. T. Thomson (Edinburgh, 1815).

83 Timothy Pont, map of Loch Tay: the head of Glen Tanar, *c.* 1583–96. National Library of Scotland, Adv.MS.70.2.9 (Pont 18).

84 Maria Steuart, 'Scots Pearls', *Scottish Historical Review*, XVII/68 (July 1920), p. 288.

85 Horace Marryat, writing of his one year's residence in Sweden, notes this in 1860. See ibid., p. 288.

86 Ibid., pp. 292–3.

87 Eddie Davies, 'Sixty Years in the River', in *The Summer Walkers: Travelling People and Pearl-Fishers in the Highlands of Scotland*, ed. Timothy Neat (Edinburgh, 2002), pp. 107–8. Victoria Finlay wrote a moving account of meeting and interviewing Eddie Davies in *Jewels: A Secret History* (New York, 2006).

88 Davies, 'Sixty Years', p. 112.

89 Neat, *Summer Walkers*, p. 227.

90 The lifecycle of *M. margaritifera* is well documented. See, for example, Ann Skinner, Mark Young and Lee Hastie, *Ecology of the Freshwater Pearl Mussel*, Conserving Natura 2000 Rivers, Ecology Series, no. 2 (Edinburgh, 2003), pp. 4–6.

91 Landman et al., *Pearls*, p. 142.

92 Finlay, *Jewels*, p. 84.

93 Author's interview with the ecologist Peter Cosgrove, 9 September 2020.

94 'Crimes Against Freshwater Pearl Mussels', Partnership for Action Against Wildlife Crime in Scotland, www.gov.scot, accessed 7 September 2020.

95 Peter Cosgrove, Lee Hastie and Iain Sime, 'Wildlife Crime and Scottish Freshwater Pearl Mussels', *British Wildlife*, XXIV (October 2012), pp. 10–13.

96 Peter Cosgrove et al., 'Recorded Natural Predation of Freshwater Pearl Mussels *Margaritifera margaritifera* (L.) in Scotland', *Journal of Conchology*, XXXIX/4 (December 2007), pp. 469–72.

97 S. J. Harrison, 'Changes in Scottish Climate', *Botanical Journal of Scotland*, XLIX/2 (1997), pp. 287–300.

98 Peter Cosgrove et al., 'Scotland's Freshwater Pearl Mussels: The Challenge of Climate Change', in *River Conservation and Management*, ed. Philip J. Boon and Paul J. Raven (Chichester, 2012), pp. 119–30.

99 In 2019 Forest Enterprise Scotland was absorbed into Forestry and Land Scotland.

100 Peter Cosgrove et al., 'Forest Management and Freshwater Pearl Mussels: A Practitioners' Perspective from the North of Scotland', *Scottish Forestry*, LXXI (2017), pp. 14–21 (p. 19).

101 Robert Burns, 'Poor Mailie's Elegy', www.scottishpoetrylibrary.org.uk, accessed 11 April 2021.

THREE Culturing Pearls, Capturing Markets, Cultivating Brands

1 Embedding pearls with chips containing large files, for example, scripture and movies, was first introduced by Galatea: Jewelry by Artist's line 'Momento', www.momentogem.com. For a technical discussion, see Artitaya Homkrajae, 'A Near-Field-Communication (NFC) Technology Device Embedded in Bead Cultured Pearls', *Gems and Gemology*, LVI/3 (Fall 2020), pp. 436–7.

2 The constant innovation in this industry can be appreciated by attending major trade shows such as the Tucson Gem and Mineral Show in Arizona. Journals such as *Gems and Gemology*, the quarterly journal of the Gemological Institute of America, present research on new technologies.

3 Most sources point to China as the site of the earliest experiments, though the date is disputed. See Neil Landman et al., *Pearls: A Natural History* (New York, 2001), p. 154.

4 Employing first-hand observers, F. Hague provides a lengthy description of the process in mid-nineteenth-century Zhejiang in 'On the Natural and Artificial Production of Pearls in China', *Journal of the Royal Asiatic Society of Great Britain and Ireland*, XVI (1856), pp. 280–84.

5 R. A. Donkin, *Beyond Price: Pearls and Pearl Fishing, Origins to the Age of Discoveries* (Philadelphia, PA, 1998), pp. 14–16; Elisabeth Strack, *Pearls* (Stuttgart, 2006), p. 113.

6 Peter Collinson, letter to Linnaeus, 30 October 1748, available online from the Linnean Society, The Linnaean Correspondence Collection, L0952, www.linnean.org, accessed 10 October 2020.

7 Linnaeus's detailed journal of this trip is preserved at the Linnean Society of London, *Iter Lapponicum* (1732), GB-110/LM/LP/TRV/1/2/1. An English translation was published after his death as *Lachesis Lapponica: A Tour in Lapland*, ed. James Edward Smith, 2 vols (London, 1811).

8 Sweden's incursions into Lapland are discussed by Gunlög Fur, *Colonialism in the Margins: Cultural Encounters in New Sweden and Lapland* (Leiden, 2006).

9 The economic context of Linnaeus's work is the subject of Lisbet Koerner's *Linnaeus: Nature and Nation* (Cambridge, 1999).

10 Linnaeus, *Lachesis Lapponica*, vol. I, pp. 207–10.

11 Ibid., p. 236.

12 Koerner, *Linnaeus: Nature and Nation*, p. 61; Fur, *Colonialism in the Margins*, pp. 51–62.

13 Linnaeus, *Lachesis Lapponica*, vol. II, p. 27.

14 Koerner in *Linnaeus: Nature and Nation*, clarifies that Linnaeus exaggerated the rigours of his Lapland trip, including the altitude of mountain passes. For example, he travelled into Norway on a well-trodden trading route.

15 Linnaeus, *Lachesis Lapponica*, vol. I, p. 265.

16 Linnaeus includes an extended description of the pearl fisheries at Purkijaur in *Lachesis Lapponica*, vol. II, pp. 103–7.

17 Koerner, *Linnaeus: Nature and Nation*, p. 61.

18 Nils Storä, 'Pearl Fishing among the Eastern Saami', *Acta Borealia*, VI/2 (1989), pp. 12–27.

19 Gemmologist and historian Jan Asplund, personal correspondence with author, 4 November 2020.

20 Storä, 'Pearl Fishing', p. 17.

21 Magnus Lagerström, letter to Linnaeus, 17 July 1748, L0930, available at www.alvin-portal.org.

22 Linnaeus, letter to François Boissier de La Croix de Sauvages, [30 June] 1748, L1003, available at www.alvin-portal.org.

23 Koerner, *Linnaeus: Nature and Nation*, p. 142.

24 Linnaeus, letter to Carl Funck, 6 February 1761, L2882, available at www.alvin-portal.org.

25 Linnaeus, letter to Funck, available at www.alvin-portal.org.

26 Linnaeus, letter to Funck; Koerner, *Linnaeus: Nature and Nation*, pp. 140–46, available at www.alvin-portal.org.

27 Linnaeus likened nature to 'an infinite larder' in a pamphlet from 1762. Quoted in Koerner, *Linnaeus: Nature and Nation*, p. 92.

28 Joseph von Rathgeb, letter to Linnaeus, 16 September 1750, L1165, available at www.alvin-portal.org.

29 Koerner, *Linnaeus: Nature and Nation*, pp. 145–6.

30 This photograph is reproduced in an early English-language biography of Mikimoto by Robert Eunson, *The Pearl King: The Story of the Fabulous Mikimoto* (Tokyo, 1964), p. 20 facing.

31 Mikimoto's humble beginnings were emphasized early in the biographical literature; in-house publicity makes much of the rags-to-riches trajectory, though his grandfather had been a successful merchant with a sea-freight

business. *Kōkichi Mikimoto Memorial Hall* (Mikimoto Pearl Island, 1994), p. 5.

32 Mari Natsuke, 'My Pearls, My Style', www.mikimoto.com, accessed 12 November 2020.

33 Jean-Baptiste Tavernier, *Travels in India*, trans. V. Ball (London, 1889), vol. II, p. 114. Landman et al. note that pearls were not commonly worn by men or women in Japan before Western fashions were introduced in the Meiji era (1868–1912), *Pearls: A Natural History*, p. 121.

34 Of the 4,158 pearls in the Shōsōin Imperial Treasure House in Nara, 3,830 were used to ornament the emperor's crown. See Shigeru Akamatsu, *Pearl Book* (Tokyo, 2015), p. 28.

35 A detailed inventory of the objects, which formed the nucleus of the Shōsōin collection, was compiled after the emperor's death. See Jiro Harada, *English Catalogue of the Treasures in the Imperial Repository Shōsōin* (Tokyo, 1932).

36 These votive offerings, buried beneath the central altar and excavated from the late nineteenth century onwards, are on public display in the treasure hall at Kōfuku-ji, Nara. Observed by author, October 2019.

37 George Kuwayama, *Far Eastern Lacquer* (Los Angeles, CA, 1978), p. 30.

38 Some of these artefacts were stored in the Shōsōin Imperial Treasure House in Nara and survive in good condition.

39 Donkin, *Beyond Price*, p. 208.

40 Kjell David Ericson, 'Nature's Helper: Mikimoto Kōkichi and the Place of Cultivation in the Twentieth Century's Pearl Empires', PhD dissertation, Princeton University, 2015, p. 53.

41 Image reproduced in Ericson, 'Nature's Helper', p. 2. I would like to acknowledge this author for drawing my attention to this photograph.

42 Quoted ibid.

43 For example, 'Japan's Pearl King Holds Rites for Oysters' Souls', *New York Times*, 31 October 1936, p. 21.

44 The chronology of Mikimoto's career is well documented. See, for instance, Kiyohito Nagai, 'A History of the Cultured Pearl Industry', *Zoological Science* (Zoological Society of Japan), XXX/10 (2013), pp. 783–93.

45 From the late eighteenth century to the nineteenth, over a hundred artificial Mount Fujis were constructed for Japanese gardens, combining artifice with devotion. See Melinda Takeuchi, 'Making Mountains: Mini-Fujis, Edo Popular Religion and Hiroshige's "One Hundred Famous Views of Edo"', *Impressions*, 24 (2002), pp. 24–47.

46 I would like to thank the marine biologists Edgar Walters, Robyn J. Crook and Bill Wright for their insightful responses to my questions about the experience of pain in bivalves.

47 K. Mikimoto, *Japanese Culture Pearls: A Successful Case of Science Applied in Aid of Nature* (Tokyo, 1907), p. 7.

48 A survey had been completed in 1892.

49 Strack, *Pearls*, p. 317.

50 Mikimoto, *Japanese Culture Pearls*, p. 14.

51 See Chapter One, 'Pearl Kingship: Shellfsh Protection and the Bayscape of Saltwater Cultivation', in Ericson, 'Nature's Helper', pp. 39–94.

52 See Barbara R. Ambros, *Bones of Contention: Animals and Religion in Modern Japan* (Honolulu, HI, 2012) for a discussion of the paradoxical attitudes to animals in Japan.

53 Both men almost certainly had knowledge of the work of William Saville-Kent, who experimented with culturing South Sea pearls in Queensland, Australia, in the early 1900s, although never wrote of his method. See Strack, *Pearls*, pp. 320–21, 470 for a discussion of this theory.

54 Akamatsu, *Pearl Book*, p. 46.

55 Mikimoto, 'Brand Story', www.mikimotoamerica.com, accessed 16 October 2020.

56 *Kōkichi Mikimoto Memorial Hall*, p. 18.

57 See Audrey Yoshiko Seo, 'Adoption, Adaptation, and Innovation: The Cultural and Aesthetic Transformations of Fashion in Modern Japan', in *Since Meiji: Perspectives on the Japanese Visual Arts, 1868–2000*, ed. J. Thomas Rimer (Honolulu, HI, 2011), pp. 471–96.

58 *Kōkichi Mikimoto Memorial Hall*, pp. 26–7.

59 'Three Minute Pearls', *Time*, 31 October 1932, p. 18.

60 George Frederick Kunz and Charles Hugh Stevenson, *The Book of the Pearl: The History, Art, Science, and Industry of the Queen of Gems* (New York, 1908), p. 293.

61 Kunz and Stevenson, *Book of the Pearl*, p. 98.

62 Léonard Rosenthal, *The Kingdom of the Pearl* (London, 1920), p. 65.

63 H. Lyster Jameson, 'The Japanese Artificially Induced Pearl', *Nature*, CVII/2691 (26 May 1921), pp. 396–8.

64 Strack, *Pearls*, p. 323.

65 Russell Shor, 'From Single Source to Global Free Market: The Transformation of the Cultured Pearl Industry', *Gems and Gemology*, XLIII/3 (Fall 2007), pp. 200–226 (p. 201).

66 The stunt was duly reported by the foreign press. See, for example, 'Three Minute Pearls', *Time*, 31 October 1932, p. 18.

67 Akamatsu, *Pearl Book*, p. 82.

68 For a discussion of the use of pearl shell in Aboriginal culture, see Kim Akerman and John E. Stanton, *Riji and Jakuli: Kimberley Pearl Shell in Aboriginal Australia* (Darwin, 1994).

69 Kimberley artist and Walmajarri elder Mumbadadi, quoted by Beau James, 'North-West Pearl Shells', Australian National Maritime Museum, www.sea.museum, accessed 5 November 2020.

70 Shor, 'From Single Source', p. 207.

71 The borders between Malaysia, Singapore, Timor Leste, Papua New Guinea and Australia were finalized between the 1870s and 1920s. See Julia Martínez and Adrian Vickers, *The Pearl Frontier: Indonesian Labor and Indigenous Encounters in Australia's Northern Trading Network* (Honolulu, HI, 2015), pp. 21–39.

72 See Stephen Mullins, *Octopus Crowd: Maritime History and the Business of Australian Pearling in Its Schooner Age* (Tuscaloosa, AL, 2019), pp. 4–5.

73 Alistair Paterson and Peter Veth, 'The Point of Pearling: Colonial Pearl Fisheries and the Historical Translocation of Aboriginal and Asian Workers

in Australia's Northwest', *Journal of Anthropological Archaeology*, LVII
(March 2020), pp. 3–4.

74 For a discussion of the role of indigenous women in this industry, see Julia
Martínez, 'Pearling Women in North Australia: Indigenous Workers and
Wives', in *Pearls, People, and Power: Pearling and Indian Ocean Worlds*,
ed. Pedro Machado et al. (Athens, OH, 2019), pp. 344–63.
75 Broome Historical Society and Museum, 'Indentured Labor',
www.broomemuseum.org.au, accessed 6 November 2020.
76 Department of Agriculture, Water and the Environment, *West Kimberley
Place Report*, undated, p. 49, available at www.awe.gov.au, accessed
10 November 2020.
77 Western Australian Museum, 'Farewell', www.museum.wa.gov.au, accessed
6 November 2020.
78 Strack, *Pearls*, p. 471.
79 Shor, 'From Single Source', p. 208.
80 For a description of the contributions of Greek immigrants and especially
the Paspalis family to the Australian South Sea Pearl Industry, see Leonard
Janiszewski and Effy Alexakis, 'White Gold, Deep Blue: Greeks in the
Australian pearling industry, 1880s–2007', in *Greek Research in Australia:
Proceedings of the Biennial International Conference of Greek Studies*,
ed. E. Close et al., Flinders University, June 2007, pp. 119–30.
81 See Stephen G. Bloom, *Tears of Mermaids: The Secret Story of Pearls*
(New York, 2009). Bloom worked a stint as a deckhand for Paspaley and
met a team of Japanese grafters. This continued influence is confirmed by
veteran pearl dealer Anil Maloo of Baggins Pearls, interview with the author,
Los Angeles, 30 October 2020.
82 Shor, 'From Single Source', pp. 208–9.
83 Author's interview with Anil Maloo. Maloo's grandfather, a pearl dealer
from Bombay, was one of the earliest Indian merchants to establish a
business in Kobe, post-Second World War.
84 Shor, 'From Single Source', p. 209.
85 Bloom describes his first-hand experience of a Hong Kong auction,
plus other distribution channels now available to producers, in *Tears
of Mermaids*.
86 Strack, *Pearls*, p. 327.
87 Shohei Shirai, *The Story of Pearls* (Tokyo, 1970), p. 51.
88 Strack, *Pearls*, p. 333.
89 In some areas of the country, wild oysters are supplemented with a small
percentage of hatchery-bred oysters, but the Australian industry largely
relies on wild stock.
90 Landman et al., *Pearls*, p. 193.
91 For a description of these operating theatres, see Bloom, *Tears of Mermaids*,
and Ben Hills, 'Pearl Jam', *Sydney Morning Herald*, 7 September 2013,
www.smh.com.au, accessed 15 November 2020.
92 Landman et al., *Pearls*, p. 167.
93 Values and volumes fluctuate constantly according to market demand.
For analysis see Clement Allan Tisdell and Bernard Poirine, 'Economics

of Pearl Farming', in *The Pearl Oyster*, ed. Paul Southgate and John Lucas (Amsterdam, 2008), pp. 473–96.
94 Bloom documents this attitude through conversations with directors of Japanese pearl companies, *Tears of Mermaids*, pp. 62, 69.
95 Monroe's ownership of Mikimoto pearls is frequently referred to in literature about the company and about pearls.
96 Strack, *Pearls*, pp. 473–4.
97 The websites for the major producers of South Sea pearls in Australia and the Philippines foreground images of water, specific farms and communities of workers. Some have branched into eco-tourism.

FOUR The Seven Pearly Sins

1 *The Pearl* was a monthly erotic magazine which ran for a brief spurt between 1879 and 1880. The medieval English poem *Pearl* dates to the late fourteenth century.
2 Delores Martinez points out the persistence of this misconception in her anthropological study of the diving village of Kuzaki, Mie Prefecture, *Identity and Ritual in a Japanese Diving Village* (Honolulu, HI, 2004).
3 Tomohiro Osaki, 'City Withdraws Controversial "Ama" Divers Mascot', *Japan Times*, 6 November 2015, www.japantimes.co.jp, accessed 10 January 2020.
4 'Wishing for pearls to send to his home in the capital', *The Manyoshu: The Nippon Gakujutsu Shinkokai Translation of One Thousand Poems*, foreword by Donald Keene (New York, 1965), p. 152.
5 For an English-language translation of this conversation see Danielle Talerico, 'Interpreting Sexual Imagery in Japanese Prints: A Fresh Approach to Hokusai's "Diver and Two Octopi"', *Impressions*, no. 23 (2001), pp. 24–41 (p. 37). Talerico unpacks the way this image would have been interpreted by its original audience and I am grateful to this scholarship for informing my discussion.
6 Ibid. Talerico argues that contemporary viewers would have drawn the connection between Hokusai's imagery and the folk tale.
7 See discussions in Timon Screech, *Sex and the Floating World: Erotic Images in Japan, 1700–1820* (Honolulu, HI, 1999).
8 Yukio Mishima, *The Sound of Waves*, trans. Meredith Weatherby (New York, 1966), p. 68.
9 A putative report documented in Hiromi Rogers, *Anjin: The Life and Times of Samurai Adams* (Folkestone, 2017), pp. 159–60.
10 Fosco Maraini, *The Island of the Fisherwomen* [1960], trans. Eric Mosbacher (New York, 1962), pp. 53, 68, 71.
11 Martinez notes manuscript evidence that Fleming was directly influenced by Maraini's photography, *Identity and Ritual*, p. 37.
12 Mishima, *Sound of Waves*, p. 138.
13 Maraini, *Island of the Fisherwomen*, p. 20.
14 Author's visit to Mikimoto Pearl Island, October 2019.
15 Plutarch, *Life of Anthony*, 28. See Prudence Jones, *Cleopatra: A Sourcebook* (Norman, OK, 2006), p. 104.

16 Pliny the Elder recounts in his *Natural History*, 9.119–21 the now famous
    tale about Cleopatra dissolving a pearl. Pliny claims the pearl is worth 10
    million sesterces, which Stacy Schiff compares to the price for a villa in her
    Cleopatra: A Life (New York, 2011), p. 122.
17 Diana E. E. Kleiner examines Cleopatra's complex relationships with her own
    family and Roman rulers in *Cleopatra and Rome* (Cambridge, MA, 2009).
18 Kara Cooney, *When Women Ruled the World* (Washington, DC, 2018),
    p. 286.
19 Cassius Dio, *Roman History*, 42.34 describes the charisma of Cleopatra,
    and her ability to captivate 'a man tired of love'. Jones, *Cleopatra*, p. 56.
20 Lucan, *On the Civil War,* Book 10, lines 139–43; Jones, *Cleopatra*, p. 68.
21 See Kleiner, *Cleopatra and Rome*, p. 21 for a discussion of Cleopatra's
    goddess status.
22 Phyllis Pray Bober details the rich variety of foods available to the ancient
    Egyptians in *Art, Culture, and Cuisine: Ancient and Medieval Gastronomy*
    (Chicago, IL, 1999).
23 R. A. Donkin, *Beyond Price: Pearls and Pearl Fishing, Origins to the Age of
    Discoveries* (Philadelphia, PA, 1998), p. 90.
24 Schiff traces the many ways Cleopatra consolidated her identification with
    Isis in *Cleopatra: A Life*, pp. 135–6, 168–9; Kleiner, *Cleopatra and Rome*,
    pp. 166–7.
25 Kleiner, *Cleopatra and Rome*, p. 100.
26 In his *Life of Antony*, 24–6, Plutarch claims that both were widely identified
    with these deities.
27 Or so Plutarch would have us believe. His *Life of Antony*, 28–9, details the
    debauchery enjoyed with Cleopatra.
28 Pliny the Elder, *Natural History*, 9.120–21, in Jones, *Cleopatra*, p. 108.
29 Several scholars have conducted experiments to dissolve pearls in vinegar,
    the most notable being Berthold Ullman, who reported his findings in
    'Cleopatra's Pearls', *Classical Journal*, LII/5 (February 1957), pp. 193–201;
    and Prudence Jones, 'Cleopatra's Cocktail', *Classical World*, CIII/2
    (Winter 2010), pp. 207–20.
30 Donkin, *Beyond Price*, p. 81.
31 Pliny the Elder, *Natural History*, 9.119, in Jones, *Cleopatra*,
    p. 107.
32 For Cleopatra's education see Schiff, *Cleopatra: A Life*, pp. 32–40; Kleiner,
    Cleopatra and Rome, pp. 72–4. For medical professionals at court, see Schiff,
    Cleopatra: A Life, p. 148.
33 The emperor Augustus donated the pearl to Venus in 25 BCE. See Marleen
    B. Flory, 'Pearls for Venus', *Historia: Zeitschrift für Alte Geschichte*, XXXVII/4
    (1988), pp. 498–504.
34 Jack Weatherford notes these statistics (and their improbability) in *Genghis
    Khan and the Making of the Modern World* (New York, 2004), p. 118.
35 Ibid., pp. xv–xvi. The banner has been lost since 1937.
36 Jack Weatherford, *Genghis Khan and the Making of the Modern World*
    (New York, 2004), p. xviii. The source gives a range of 11–12 million
    contiguous sq. miles.

37 *The Secret History of the Mongols: A Mongolian Epic Chronicle of the Thirteenth Century*, trans. Igor de Rachewiltz, shorter version ed. John C. Street, University of Wisconsin, Madison, Books and Monographs, Book 4, sections 76–8, available at https://cedar.wwu.edu, accessed 10 March 2020.

38 Ata-Malik Juvaini, *The History of the World Conqueror*, trans. John Andrew Boyle (Cambridge, MA, 1958), p. 129. This episode is related by Thomas Allsen in *The Steppe and the Sea: Pearls in the Mongol Empire* (Philadelphia, PA, 2019), p. 23.

39 Ivan Bolotov et al., 'Multi-Locus Fossil-Calibrated Phylogeny, Biogeography and a Subgeneric Revision of the Margaritiferidae (Mollusca: Bivalvia: Unionoida)', *Molecular Phylogenetics and Evolution*, CII (October 2016), pp. 104–21, DOI: 10.1016/j.ympev.2016.07.020; author's correspondence with Bolotov, 24 March 2020.

40 Jack Weatherford discusses the role of animism in Genghis Khan's worldview in *Genghis Khan and the Quest for God: How the World's Greatest Conqueror Gave us Religious Freedom* (New York, 2016).

41 The mountain and surrounding sacred landscape were added to UNESCO's World Heritage List in 2015. See https://whc.unesco.org, accessed 14 November 2021.

42 Allsen, *The Steppe and the Sea*, p. 24.

43 Juzjani cites eyewitness accounts from an envoy of Khwarazmian ambassadors who visited the city after its fall. See Peter Jackson, *The Mongols and the Islamic World: From Conquest to Conversion* (New Haven, CT, 2017), p. 71.

44 Allsen, *The Steppe and the Sea*, pp. 59–61.

45 Quoted by Weatherford, *Genghis Khan and the Quest for God*, p. 200.

46 See Allsen, *The Steppe and the Sea*, pp. 69–71.

47 Suetonius, 'The Life of Caligula', in *The Lives of the Twelve Caesars*, trans. J. C. Rolfe (London, 1913), www.penelope.uchicago.edu, accessed 20 April 2020.

48 For a discussion of the excess of the Roman table, see Christine Baumgarthuber, 'Dinner with Caligula', *New Inquiry*, 17 May 2016, www.thenewinquiry.com, accessed 7 April 2020.

49 Suetonius, 'Caligula', 37.

50 Pliny the Elder, *Natural History*, 37.6; 9.58, www.perseus.tufts.edu, accessed 12 April 2020.

51 Suetonius, 'Caligula', 55.

52 Ibid., 46.

53 J.G.F. Hind, 'Caligula and the Spoils of Ocean: A Rush for Riches in the Far North-West?', *Britannia*, XXXIV (2003), pp. 272–4.

54 Suetonius claimed that Julius Caesar had invaded Britain hoping for pearls: Suetonius, 'The Life of Julius Caesar', in *The Lives of the Twelve Caesars*, trans. J. C. Rolfe (London, 1913), 47.1, www.penelope.uchicago.edu, accessed 30 April 2020.

55 See, for example, Aloys Winterling et al., *Caligula: A Biography* (Berkeley, CA, 2011).

56 Pliny the Elder, *Natural History*, 37.6.
57 The Stables Treasury in the Moscow Kremlin Museums is one example of a unique museum collection of historical horse tack with a wide range of jewelled embellishments.
58 John Knox, *The Works of John Knox*, ed. John Laing (Edinburgh, 1848), vol. II, p. 267.
59 A 1561 inventory in French lists the jewels that belonged to Mary, Queen of Scots, after the death of her first husband, King Francis II. Joseph Robertson, ed., *Inventaires de la Royne Descosse Douairiere de France: Catalogues of the jewels, dresses, furniture, books, and paintings of Mary Queen of Scots, 1556–1569* (Edinburgh, 1863), pp. 7–17.
60 Michael Pearce, 'Edinburgh Castle, 1567–1563: The Jewels Mary Queen of Scots Left Behind', unpublished research report for Historic Environment Scotland, 2016, p. 40.
61 Quoted in Kate Williams, *The Betrayal of Mary, Queen of Scots* (London, 2018), p. 8.
62 Steven Thiry, '"In Open Shew to the World": Mary Stuart's Armorial Claim to the English Throne and Anglo-French Relations (1559–1561)', *English Historical Review*, CXXXII/559 (December 2017), pp. 1405–39.
63 Pearce, 'Edinburgh Castle', p. 2.
64 Ibid., p. 43.
65 A gold crown, in Shakespeare's time, was worth a quarter of a pound, so 16,000 crowns was worth £4,000. Taking into account inflation since 1568, this sum represents over £2.1 million. See www.bankofengland.co.uk, accessed 14 November 2021. J. Duncan Mackie, 'Queen Mary's Jewels', *Scottish Historical Review*, XVIII/70 (January 1921), pp. 93–6; Pearce, 'Edinburgh Castle', pp. 40–41.
66 For the conflicting public and private motivations in the English and Scottish courts to Mary's execution, see Susan Doran, 'Revenge Her Foul and Most Unnatural Murder? The Impact of Mary Stewart's Execution on Anglo-Scottish Relations', *History*, LXXXV/280 (October 2000), pp. 589–612.
67 Williams, *The Betrayal of Mary*, p. 331.
68 Marion McCready, 'Mary Stuart', *Poetry*, CCVIII/3 (June 2016), p. 63.
69 Steinbeck and Ricketts published *Sea of Cortez* in 1941. They were accompanied on their trip by a small crew and Steinbeck's first wife, Carol, who is absent from the book. An expanded version with Steinbeck's tribute to Ricketts, who had died in 1948, was released in 1951 as *The Log from the Sea of Cortez*.
70 John Steinbeck, *The Log from the Sea of Cortez* [1951] (London, 1986), pp. 86–7.
71 Andrea Saenz-Arroyo et al., 'The Value of Evidence about Past Abundance: Marine Fauna of the Gulf of California through the Eyes of 16th to 19th Century Travellers', *Fish and Fisheries*, VII (2000), pp. 128–46 (p. 139).
72 Mario Monteforte and Micheline Cariño-Olvera, 'A History of Nacre and Pearls in the Gulf of California', in *Coastal Heritage and Cultural Resilience*, ed. Lisa L. Price and Nemer E. Narchi (New York, 2018), pp. 79–112 (p. 86).

73 For an introduction to and English translation of Sebastián Vizcaíno's narrative, see H. R. Wagner, 'Pearl Fishing Enterprises in the Gulf of California', *Hispanic American Historical Review*, x/2 (May 1930), pp. 188–218.

74 The reports submitted by these explorers stress the hostility of the terrain, paucity of pearls and problems with supplies. An account by Father Cavallero Carranco of the voyage made by Captain Francisco de Lucenilla is one of the most detailed. Juan Cavallero Carranco, *The Pearl Hunters in the Gulf of California 1668: Summary Report of the Voyage Made to the Californias By Captain Francisco De Lucenilla*, trans. W. Michael Mathes (Los Angeles, CA, 1966).

75 Steinbeck, *Log*, p. 94.

76 That California was a peninsula was known by the mid-sixteenth century, but European maps persisted in depicting it as an island well into the eighteenth century. See Justin Hyland, 'Debunking the Myth: Jesuit Texts and History and Archeology in Baja California', *Kroeber Society Anthropology Papers*, LXXIX (1995), pp. 177–88.

77 Monteforte and Cariño-Olvera, 'A History of Nacre', p. 89.

78 Melinda Knight, 'Historical Context of *The Pearl*: Steinbeck's Vision of Mexico', in *Critical Insights: The Pearl, by John Steinbeck*, ed. Laura Nicosia and James F. Nicosia (Hackensack, NJ, 2019), p. 6.

79 Steinbeck, *Log*, p. 119.

80 Steinbeck relates the folktale ibid., p. 120.

81 John Steinbeck, *The Pearl* [1947] (New York, 1992), p. 16.

82 Ibid., p. 53.

83 Gonzalo de Francia, 'Memorial of Gonzalo de Francia' [1596], trans. and published in *Hispanic American Historical Review*, x/2 (May 1930), p. 220.

84 For an analysis of linguistically distinct groups, see William C. Massey, 'Tribes and Languages of Baja California', *Southwestern Journal of Anthropology*, v/3 (Autumn 1949), pp. 272–307.

85 Mario Monteforte and Micheline Cariño, 'Episodes of Environmental History in the Gulf of California: Fisheries, Commerce, and Aquaculture of Nacre and Pearls', in *A Land Between Waters: Environmental Histories of Modern Mexico*, ed. Christopher R. Boyer (Tucson, AZ, 2012), pp. 245–76 (pp. 251–2).

86 Quoted in Molly Warsh, *American Baroque: Pearls and the Nature of Empire, 1492–1700* (Chapel Hill, NC, 2018), p. 146.

87 Kevin Dawson, *Undercurrents of Power: Aquatic Culture in the African Diaspora* (Philadelphia, PA, 2018), p. 69.

88 Peter Gerhard, 'Pearl Diving in Lower California, 1533–1830', *Pacific Historical Review*, xxv/3 (August 1956), pp. 239–49.

89 Robert Hardy, *Travels in the Interior of Mexico in 1825, 1827 and 1828* (London, 1829), pp. 256, 449.

90 Ibid., p. 250.

91 Ibid., pp. 251–2.

92 Ibid., pp. 252–3.

93 Sanford A. Mosk, 'Capitalistic Development in the Lower California Pearl Fisheries', *Pacific Historical Review*, x/4 (December 1941), pp. 461–8 (p. 465).

94 Monteforte and Cariño, 'A History of Nacre', p. 92.
95 Paul Southgate et al., 'Exploitation and Culture of Major Commercial Species', in *The Pearl Oyster*, ed. Paul Southgate and John Lucas (Amsterdam, 2008), pp. 303–55 (p. 342).
96 Robert H. Jackson, 'Epidemic Disease and Population Decline in the Baja California Missions, 1697–1834', *Southern California Quarterly*, LXIII/4 (Winter 1981), pp. 308–46 (p. 308).
97 Massey, 'Tribes and Languages of Baja California', p. 272.

FIVE And the Seven Virtues

1 Dante, *Paradiso*, 2.34–6, trans. Allen Mandelbaum in 'Digital Dante', https://digitaldante.columbia.edu, accessed 28 August 2021.
2 *Pearl*, trans. Simon Armitage (New York, 2016), p. 45.
3 Ibid., p. 19.
4 Ibid., pp. 95, 99.
5 Pedro E. Saucedo and Paul C. Southgate, 'Reproduction, Development and Growth', in *The Pearl Oyster*, ed. Paul Southgate and John Lucas (Amsterdam, 2008), pp. 131–86.
6 For the persistence of this idea see R. A. Donkin, *Beyond Price: Pearls and Pearl Fishing, Origins to the Age of Discoveries* (Philadelphia, PA, 1998), pp. 1–14.
7 For a fuller discussion see Catherine L. Howey, 'Dressing a Virgin Queen: Court Women, Dress, and Fashioning the Image of England's Queen Elizabeth I', *Early Modern Women*, IV (Fall 2009), pp. 201–8.
8 Ki Hackney and Diana Edkins, *People and Pearls: The Magic Endures* (New York, 2000), pp. 48–51.
9 Walter Raleigh, 'The Ocean to Cynthia', in *The Poems of Sir Walter Raleigh: Collected And Authenticated, with Those of Sir Henry Wotton and Other Courtly Poets from 1540 to 1650*, ed. J. Hannah (London, 1892), p. 34.
10 Meryl Bailey teases out these meanings in 'Salvatrix Mundi: Representing Queen Elizabeth I as a Christ Type', *Studies in Iconography*, XXIX (2008), pp. 176–215.
11 *Pearl*, pp. 65–7.
12 Quoted in E. de Jongh, 'Pearls of Virtue and Pearls of Vice', *Simiolus: Netherlands Quarterly for the History of Art*, VIII/2 (1975–6), p. 76.
13 *Pearl*, p. 151.
14 Ibid., p. 97.
15 The line is, of course, the translation by poet Simon Armitage. The original is just as alliterative: 'Thyn angel-hauyng so clene cortez', ibid., pp. 96–7.
16 Author's correspondence with Lauren Gill, archaeologist at Glasgow University, 2 February 2021. While the scapula is not an ideal bone for osteo-archaeological study, its tiny size does accord with written records of Queen Margaret's addiction to fasting.
17 Inventory of Philip II of holy relics in Escorial, 1571–8. See Catherine Keene, *Saint Margaret, Queen of the Scots: A Life in Perspective* (New York, 2013), p. 134.

18 Turgot, *Vita Margaretae* (*c.* 1104), in *Ancient Lives of Scottish Saints*, trans. W. M. Metcalfe (Paisley, 1895), p. 299.

19 Ibid., p. 302.

20 Ibid., pp. 304–5; see also Joanna Huntington, 'St Margaret of Scotland: Conspicuous Consumption, Genealogical Inheritance, and Post-Conquest Authority', *Journal of Scottish Historical Studies*, XXXIII/2 (November 1913), pp. 149–64.

21 Turgot, *Vita*, p. 314. Keene argues that Turgot may have exaggerated but would not have fabricated the details of Margaret's asceticism, *Saint Margaret*, p. 71.

22 Sangdong Lee, 'The Miracles and Cult of St Margaret of Scotland', *Scottish Historical Review*, XCVII/1 (April 2018), pp. 1–11.

23 Quoted in Keene, *Saint Margaret*, p. 120. Recording the fragrance of a saint's remains was a hagiographical trope. See also Kate Ash, 'St Margaret and the Literary Politics of Scottish Sainthood', in *Sanctity as Literature in Late Medieval Britain*, ed. Eva Von Contzen and Anke Bernau (Manchester, 2015), pp. 18–37 (p. 25).

24 Lee, 'Miracles and Cult of St Margaret', pp. 2–3.

25 For the dispersal of Margaret's relics, see Keene, *Saint Margaret*, pp. 133–4.

26 Turgot, *Vita*, p. 299.

27 Ariane Fennetaux, 'Fashioning Death/Gendering Sentiment: Mourning Jewelry in Britain in the Eighteenth Century', in *Women and the Material Culture of Death*, ed. Maureen Daly Goggin and Beth Fowkes Tobin (Abingdon, 2014), pp. 27–50 (p. 29).

28 Quoted ibid.

29 Donkin, *Beyond Price*, pp. 7, 155; George Frederick Kunz and Charles Hugh Stevenson, *The Book of the Pearl: The History, Art, Science, and Industry of the Queen of Gems* (New York, 1908), p. 35.

30 Kunz and Stevenson, *Book of the Pearl*, p. 307.

31 Kunz and Stevenson describe the pearl jewellery auctioned at this sale, ibid., p. 470.

32 Hackney and Edkins, *People and Pearls*, p. 68; Elizabeth Strack, *Pearls* (Stuttgart, 2006), p. 301.

33 Camilla Apcar, 'The Empress Eugenie Pearls', *Financial Times*, 5 June 2015, p. 13.

34 '*En mi tiempo las perlas significaban lágrimas.*' Federico García Lorca, *La Casa de Bernarda Alba*, Act III (1936), www.cervantesvirtual.com, accessed 30 April 2021. I am grateful to Natalie Lawler for drawing my attention to this source.

35 Fennetaux, 'Fashioning Death', pp. 72–3.

36 Quoted by Deborah Lutz, 'The Dead Still Among Us: Victorian Secular Relics, Hair Jewelry, and Death Culture', *Victorian Literature and Culture*, XXXIX/1 (2011), pp. 127–42 (p. 132).

37 For biographical details, see Paul S. Harris, 'Gilbert Stuart and a Portrait of Mrs Sarah Apthorp Morton', *Winterthur Portfolio*, I (1964), pp. 198–220; Walter Muir Whitehill, 'Perez Morton's Daughter Revisits Boston in 1825', *Proceedings of the Massachusetts Historical Society*, LXXXII (1969), pp. 21–47.

38 Sarah Wentworth Morton, introduction to 'Ouâbi, Or the Virtues
    of Nature: An Indian Tale', Evans Early American Imprint Edition,
    https://quod.lib.umich.edu, accessed 29 August 2021.
39 Morton, 'Ouâbi', canto 1.
40 Angela Vietto, *Women and Authorship in Revolutionary America*
    (Abingdon, 2016), p. 83.
41 Discussed by Amy Elizabeth Shoultz, 'A Revolutionary Idea: Gilbert Stuart
    Paints Sarah Morton as the First Woman of Ideas in American Art', PhD
    dissertation, The University of Texas at Austin, 2008.
42 'Memento' was published in Sarah Wentworth Morton, *My Mind and Its
    Thoughts, in Sketches, Fragments, and Essays* (Boston, MA, 1823), pp. 255–6.
43 Sarah Wentworth Morton, 'Beacon Hill', Apology for the Poem, Evans
    Early American Imprint Edition, https://quod.lib.umich.edu, accessed
    29 August 2021.
44 Ibid.
45 Harris, 'Gilbert Stuart', p. 220.
46 Gilbert Stuart, lines to Sarah Wentworth Morton published in the *Port Folio*
    18 June 1803, p. 185, quoted in George Mason, *The Life and Works of Gilbert
    Stuart* (New York, 1894), p. 228.
47 Chanel has been the subject of numerous biographies, including the
    meticulous Edmonde Charles-Roux, *Chanel*, trans. Nancy Amphoux
    (London, 1976). Paul Morand based his *Allure of Chanel* [1976], trans. Euan
    Cameron (London, 2013), on conversations with the designer in 1946. More
    recently, Rhonda Garelick's *Mademoiselle: Coco Chanel and the Pulse of
    History* (New York, 2014) provides an authoritative account of the designer
    and her historical context.
48 Quoted in Amy de la Haye and Shelley Tobin, *Chanel: The Couturiere at
    Work* (New York, 1996), p. 19.
49 Quoted in Justine Picardie, *Coco Chanel: The Legend and the Life*
    (London, 2010), p. 197.
50 The emperor Jahangir recorded the birth of this grandson in his memoir,
    The Jahangirnama: Memoirs of Jahangir, Emperor of India, trans., ed.
    and annotated by Wheeler M. Thackston (Oxford, 1999), p. 172.
51 See Elaine Wright, 'Mughal Portraiture and Drawing', in Elaine Wright
    et al., *Muraqqa': Imperial Mughal Albums from the Chester Beatty Library,
    Dublin* (Alexandria, VA, 2008), pp. 164–77.
52 Neil Landman et al. analyse pearl size in relation to the iris of a human eye,
    and suggests *P. maxima* as the probable source, *Pearls: A Natural History*
    (New York, 2001), pp. 108–12.
53 For an overview of the importance of giving in Islamic courts, see Linda
    Komaroff et al., *Gifts of the Sultan: The Arts of Giving at the Islamic Courts*
    (New Haven, CT, 2011).
54 One of the best-known and detailed records is the *Book of Gifts and Rarities*,
    a ninth- through fifteenth-century manuscript edited in the fifteenth century
    by Muhammad Hamidullah. For the record of the gift to Khosrow I, see *Book
    of Gifts and Rarities*, trans. with commentary by Ghada Hijjawi Qaddumi
    (Cambridge, MA, 1996), p. 62.

55 Ibid., p. 68.

56 Ibid., pp. 126–7.

57 The weighing on solar and lunar birthdays was borrowed from a royal Hindu tradition. See Susan Stronge, 'Imperial Gifts at the Court of Hindustan', in *Gifts of the Sultan: The Arts of Giving at the Islamic Courts* (Los Angeles, CA, and New Haven, CT, 2011), pp. 171–83 (p. 176).

58 *The Jahangirnama* records many such occasions. See also ibid., pp. 171–83.

59 Ibid., p. 177.

60 Ibid., p. 171.

61 Aurangzeb (Emperor Alamgir) has a complicated legacy and has been reviled for religious intolerance. Audrey Truschke aims to give a more nuanced reading of his life in *Aurangzeb: The Life and Legacy of India's Most Controversial King* (Stanford, CA, 2017).

62 This famous heist has been the subject of several books, the earliest by the barrister Christmas Humphreys, *The Great Pearl Robbery of 1913: A Record of Fact* (London, 1929).

63 The following account of the robbery is constructed from reports in the *North China Herald*, 14 December 1918 and 8 March 1919.

64 Biographical information compiled from record for James Cruickshank in China Families, a platform directed by Robert Bickers, University of Bristol, www.chinafamilies.net, accessed 10 November 2020.

65 University of Bristol Historical Photographs of China ref. A104-144, www.hpcbristol.net, accessed 25 January 2021.

66 For Shanghai history see Lynn Pan, *Old Shanghai: Gangsters in Paradise* (Singapore, 2011); Frances Wood, *No Dogs and Not Many Chinese: Treaty Port Life in China, 1843–1943* (London, 1998).

SIX Embodied

1 This appraisal occurred in India shortly after its independence in 1947, and is recounted in Lois Sherr Dubin, *The History of Beads from 100,000 BC to the Present* (New York, 2009), p. 306.

2 W. M. Flinders Petrie, *The Hawara Portfolio: Paintings of the Roman Age* (London, 1913), no. 50.

3 These portraits have been the subject of extensive scholarly and scientific interest. The 'Ancient Panel Paintings: Examination, Analysis, and Research Project (APPEAR)', initiated in 2013 by Getty conservator Marie Svoboda, aims to consolidate technical knowledge of the 1,028 extant portraits in collections worldwide. Of 324 total entries in the database, there are 69 pearls depicted on female portraits. My thanks to Marie Svoboda for this data.

4 Pliny the Elder, *The Natural History*, 9.56. ed. John Bostock and H. T. Riley, in the Perseus Digital Library, www.perseus.tufts.edu, accessed 5 May 2020.

5 For example, a pair of gold earrings with pendant pearls in the British Museum, 1917,0601.2582.

6 Toby Wilkinson, *The Nile: A Journey Downriver Through Egypt's Past and Present* (New York, 2014), p. 267.

7   See Katia Schörle, 'Pearls, Power, and Profit: Mercantile Networks and Economic Considerations of the Pearl Trade in the Roman Empire', in *Across the Ocean: Nine Essays on Indo-Mediterranean Trade*, ed. Federico De Romanis and Marco Maiuro (Boston, MA, 2015), pp. 43–54.

8   Donkin points to their absence in the archaeological and written record, but proposes they are unlikely to have been unknown by a culture that prized gems so highly. R. A. Donkin, *Beyond Price: Pearls and Pearl Fishing, Origins to the Age of Discoveries* (Philadelphia, PA, 1998), pp. 42–4.

9   The Stockholm Papyrus, a Graeco-Roman collection of recipes recovered near Thebes, provides a formula for making fake pearls, National Library of Sweden, Acc 2013/75, olim Dep.45.

10 Caroline Cartwright et al., 'Portrait Mummies from Roman Egypt: Ongoing Collaborative Research on Wood Identification', *British Museum Technical Research Bulletin*, V (2011), pp. 49–58.

11 Johanna Salvant et al., 'A Roman Egyptian Painting Workshop: Technical Investigation of the Portraits from Tebtunis, Egypt', *Archaeometry*, LX (2018), pp. 815–33.

12 Ibid., p. 10.

13 Although there is ongoing debate whether mummy portraits were acquired by living patrons and then repurposed at the time of death, when mummy portraits are compared to CT scans of the mummified remains, the dates match. See, for instance, Susan Walker, 'Mummy Portraits and Roman Portraiture', in *Ancient Faces: Mummy Portraits from Roman Egypt*, ed. Susan Walker (New York, 2000), pp. 23–5 (p. 24).

14 For further information on red shroud mummies, see Lorelei H. Corcoran and Marie Svoboda, *Herakleides: A Portrait Mummy from Roman Egypt* (Los Angeles, CA, 2011).

15 For a discussion on funerary practice and belief in Roman Egypt, see Christina Riggs, *The Beautiful Burial in Roman Egypt: Art, Identity, and Funerary Religion* (Oxford, 2006).

16 Scanning of complete mummies has revealed that in many cases portraits conformed to the features of the corpse.

17 Umberto Pappalardo discusses the worship of Isis in Pompeii in *The Splendour of Roman Wall Painting* (Los Angeles, CA, 2008), p. 15.

18 Euphrosyne Doxiadis notes the identification of the female mummy with Isis in *The Mysterious Fayum Portraits: Faces from Ancient Egypt* (New York, 1995), p. 39.

19 As a court artist retained by the Gonzaga, Mantegna likely painted this *Adoration* for a family member, or at their behest for one of their circle.

20 This extraordinary item of jewellery appeared in American *Vogue*, December 1920. Reproduced in Hans Nadelhoffer, *Cartier: Jewelers Extraordinary* (New York, 1984), p. 139.

21 Dawson W. Carr, *Andrea Mantegna: The Adoration of the Magi* (Los Angeles, CA, 1997), p. 52.

22 Tertullian, *Adversus Marcionem*, 3:13, trans. Ernest Evans (Oxford, 1972). Tertullian was the first writer to claim the magi were foreign kings.

23 Carr, *Andrea Mantegna*, p. 56.

24 Slavery (though not slave trading) had no legal status in the Low Countries, England and Sweden. For a discussion of the status of African slaves in Europe see Kate Lowe, 'The Lives of African Slaves and People of African Descent in Renaissance Europe', in *Revealing the African Presence in Renaissance Europe*, ed. Joaneath Spicer (Baltimore, MD, 2012), pp. 13–34.

25 For example, a black retainer is depicted by Benozzo Gozzoli in *The Journey of the Magi*, 1459–61, Medici Riccardi Palace, Florence.

26 Kate Lowe, 'Black Gondoliers and Other Black Africans in Renaissance Venice', *Renaissance Quarterly*, LXVI/2 (Summer 2013), pp. 41–52.

27 Kate Lowe, 'Visual Representations of an Elite: African Ambassadors and Rulers in Renaissance Europe', in *Revealing the African Presence*, pp. 99–115.

28 Joaneath Spicer, 'European Perceptions of Blackness as Reflected in the Visual Arts', in *Revealing the African Presence*, pp. 35–59 (p. 43). See also Paul Kaplan, 'Isabella d'Este and Black African Women', in *Black Africans in Renaissance Europe*, ed. T. F. Earle and K. P. Lowe (Cambridge, 2005), pp. 125–54.

29 While there were sizeable numbers of known embassies from Africa, their visual documentation is more slender. Lowe traces the representation of African ambassadors, diplomats and rulers in 'Visual Representations', pp. 99–115. She underscores that portraiture of African elites in this period is unstable, and apt to slip into genres such as the black magus.

30 The portrait is in the Kunsthistorisches Museum, Vienna, inventory number Gemäldegalerie, 83.

31 Roger Jones, 'Mantegna and Materials', *I Tatti Studies in the Italian Renaissance*, II (1987), pp. 71–90 (p. 73).

32 Andrea Rothe, 'Andrea Mantegna's *Adoration of the Magi*', in *Historical Painting Techniques, Materials, and Studio Practice*, Preprints of a Symposium, University of Leiden, the Netherlands, 2629 June 1995, ed. Arie Wallert et al. (Los Angeles, CA, 1995), pp. 111–16.

33 There were, though, pearl fisheries in the Red Sea and along the coast of East Africa, as described by Donkin, *Beyond Price*, pp. 119–22.

34 Molly Warsh, 'A Political Ecology in the Early Spanish Caribbean', *William and Mary Quarterly*, LXXI/4 (October 2014), pp. 517–48 (p. 525).

35 A basic Internet search turns up endless results in print and on websites and museum databases.

36 E. de Jongh discusses literary references in the Dutch Republic in 'Pearls of Virtue and Pearls of Vice', *Simiolus: Netherlands Quarterly for the History of Art*, VIII/2 (1975–6), pp. 69–97.

37 Molly Warsh, *American Baroque: Pearls and the Nature of Empire, 1492–1700* (Chapel Hill, NC, 2018), pp. 233–7.

38 Abbie Vandivere, a paintings conservator at the Mauritshuis, documented the findings in 'Girl with a Blog', www.mauritshuis.nl/en, accessed 6 June 2020.

39 The portrait was identified variously as *Head of a Girl* and *Girl with a Turban* in art-historical literature for most of the twentieth century. The Mauritshuis, where it is now held, titles it *Meisje met de parel*.

40 Jørgen Wadum et al., *Vermeer Illuminated: Conservation, Restoration and Research: A Report on the Restoration of the View of Delft and the Girl with a Pearl Earring by Johannes Vermeer* (The Hague, 1994), p. 18.

41 For a detailed analysis of the pigments used in this work see John K. Delaney et al., 'Mapping the Pigment Distribution of Vermeer's *Girl with a Pearl Earring*', *Heritage Science*, VIII/4 (2020), doi.org/10.1186/s40494-019-0348-9.

42 Annelies van Loon et al., 'Beauty is Skin Deep: The Skin Tones of Vermeer's *Girl with a Pearl Earring*', *Heritage Science*, VII/102 (2019), doi.org/10.1186/s40494-019-0344-0.

43 S. De Meyer et al., 'Macroscopic X-ray Powder Diffraction Imaging Reveals Vermeer's Discriminating Use of Lead White Pigments in *Girl with a Pearl Earring*', *Science Advances*, V/8 (August 2019), doi: 10.1126/sciadv.aax1975.

44 This becomes apparent under magnification.

45 For a history of artificial pearls, see Charlotte Eng and Maria Fusco, 'Fish Scales and Faux Pearls: A Brief Exploration into the History of Manufacturing Faux Pearls', *Textile History*, XLIII/2 (2012), pp. 250–59.

46 Arthur K. Wheelock Jr, 'Johannes Vermeer, Woman Holding a Balance, *c.* 1664', *Dutch Paintings of the Seventeenth Century*, National Gallery of Art, www.nga.gov, accessed 4 June 2020.

47 Quoted in Carla Petievich, 'Innovations Pious and Impious: Expressive Culture in Nawabi Lucknow', in *India's Fabled City: The Art of Courtly Lucknow*, ed. Stephen Markel (Los Angeles, CA, 2011), pp. 103–19 (p. 109).

48 Quoted in Muzaffar Alam and Sanjay Subrahmanyam, 'Of Princes and Poets in Eighteenth-Century Lucknow', in *India's Fabled City: The Art of Courtly Lucknow*, ed. Stephen Markel (Los Angeles, CA, 2011), pp. 187–97 (p. 191).

49 Mir Hasan wrote extensively of the delights of Faizabad. See Alam and Subrahmanyam, 'Of Princes and Poets', p. 191.

50 Donkin, *Beyond Price*, p. 169.

51 For a discussion on the development of painting in Awadh in this period, see Malini Roy, 'Origins of the Late Mughal Painting Tradition in Awadh', in *India's Fabled City: The Art of Courtly Lucknow*, ed. Stephen Markel (Los Angeles, CA, 2011), pp. 165–86.

52 Puneeta Sharma, 'Paper, Pigments and Pearls', *Icon News*, 63 (March 2016), pp. 35–6.

53 Mir Hasan, *Masnawiyat-i Hasan*, ed. Wahid Qureshi (Lahore, 1966), p. 190, quoted in Alam and Subrahmanyam, p. 190.

54 Monika Kopplin, ed., *Lacquerware in Asia, Today and Yesterday* (Paris, 2002), p. 73.

55 See, for instance, objects excavated from Ur in the British Museum.

56 Unknown author, *Papier Mâché: With Clear Directions for Decorating and Inlaying It With Pearl* (London, n.d.).

57 Kopplin, *Lacquerware in Asia*, p. 42.

58 Quoted ibid., p. 43.

59 Günther Heckmann, *Urushi No Waza: Japanese Lacquer Technology* (Ellwangen, 2002), pp. 117–21, 159–61.

60 This project was recorded as a photographic exhibition and as a book. See Catherine Opie et al., *700 Nimes Road* (New York, 2015).

61 Ibid., p. 128.

62 Ibid.

63 Opie, '"I Do Like to Stare": Catherine Opie on Her Portraits of Modern America', interview, NPR, 3 February 2016, www.npr.org, accessed 30 March 2021.

64 This provenance has been called into question; the diamond was at the heart of a legal dispute in 2015 between Christie's and Elizabeth Taylor's estate.

65 Hilton Als, staff writer at the *New Yorker*, suggested that the rooms and objects were 'totems to aura' in the essay he contributed to Opie's publication. Hilton Als, 'Elizabeth and Cathy', *700 Nimes Road*, p. 11.

66 Christie's, 'Post-Sale Release: The Most Valuable Private Collection of Jewels Sold at Auction', 14 December 2011, www.christies.com, accessed 29 June 2020.

67 George Frederick Kunz and Charles Hugh Stevenson note the many discrepancies about La Peregrina's provenance, *The Book of the Pearl: The History, Art, Science, and Industry of the Queen of Gems* (New York, 1908), p. 452. See, too, Landman et al., *Pearls*, p. 20, and Elizabeth Strack, *Pearls* (Stuttgart, 2006), p. 300. The blogosphere is rife with sites compounding the confusion.

68 Landman et al., *Pearls,* p. 20.

69 Donkin, *Beyond Price*, pp. 311–12.

70 Landman et al., *Pearls*, p. 20.

71 Kunz and Stevenson, *Book of the Pearl*, p. 452.

72 Louis de Rouvroy, duc de Saint-Simon, *Mémoires Complets et Authentiques du Duc de Saint-Simon sur le Siècle de Louis XIV et la Régence*, vol. XIX [1722] (Paris, 1841), vol. XXXVII, p. 140, excerpted and translated by Léonard Rosenthal in *The Kingdom of the Pearl* (London, 1920), pp. 87–8.

73 Elizabeth Taylor, *My Love Affair with Jewelry* (London, 2003), pp. 90–91.

74 Opie, *700 Nimes Road*, p. 128.

Bibliography

Akerman, Kim, and John E. Stanton, *Riji and Jakuli: Kimberley Pearl Shell in Aboriginal Australia* (Darwin, 1994)

Allsen, Thomas, *The Steppe and the Sea: Pearls in the Mongol Empire* (Philadelphia, PA, 2019)

Anscombe, Frederick, *The Ottoman Gulf: The Creation of Kuwait, Saudi Arabia and Qatar* (New York, 1997)

Aquilina, Berni, and William Reed, *Lure of the Pearl: Pearl Culture in Australia* (Broome, 1997)

Aschmann, Homer, ed., *The Natural and Human History of Baja California, from Manuscripts by Jesuit Missionaries* (Los Angeles, CA, 1966)

Bach, J.P.S., *The Pearling Industry of Australia: An Account of Its Social and Economic Development* (Canberra, 1955)

Banning, George Hugh, *In Mexican Waters* (London, 1925)

Bari, Hubert, and David Lam, *Pearls* (Milan, 2009)

—, and Susan Hendrickson, *The Pink Pearl: A Natural Treasure of the Caribbean* (Milan, 2007)

Barnett, Cynthia, *The Sound of the Sea: Seashells and the Fate of the Oceans* (New York, 2021)

Becker, Vivienne, *The Pearl Necklace* (New York, 2016)

Benham, Clarence, *Diver's Luck: A Story of Pearling Days* (Sydney, 1949)

Bernstein, Beth, *If These Jewels Could Talk: The Legends behind Celebrity Gems* (Woodbridge, Suffolk, 2015)

Al-Bīrūnī, *The Book Most Comprehensive in Knowledge on Precious Stones*, trans. Fritz Krenkow, ed. Hakjim Mohammad Said (Islamabad, 1989) Akamatsu, Shigeru, *Pearl Book* (Tokyo, 2015)

Bloom, Stephen G., *Tears of Mermaids: The Secret Story of Pearls* (New York, 2009)

Boettcher, Graham C., ed., *The Look of Love: Eye Miniatures from the Skier Collection* (Birmingham, AL, 2012)

Book of Gifts and Rarities, trans. with commentary by Ghada Hijjawi Qaddumi (Cambridge, MA, 1996)

Bose, Sugata, *A Hundred Horizons: The Indian Ocean in the Age of Global Empire* (Cambridge, MA, 2006)

Boyer, Christopher R., ed., *A Land Between Waters: Environmental Histories of Modern Mexico* (Tucson, AZ, 2012)

Burdett, Anita L. P., ed., *Records of the Persian Gulf Pearl Fisheries, 1857–1962*, 4 vols (Cambridge, 1995)

Butcher, John G., *The Closing of the Frontier: A History of the Marine Fisheries of Southeast Asia, c. 1850–2000* (Singapore, 2004)

Cahn, Alvin Robert, *Pearl Culture in Japan* (Tokyo, 1949)

Campbell Pedersen, Maggie, *Gem and Ornamental Materials of Organic Origin* (Oxford, 2004)

Carter, Robert A., *Sea of Pearls: Seven Thousand Years of the Industry that Shaped the Gulf* (London, 2012)

Cavallero Carranco, Juan, *The Pearl Hunters in the Gulf of California 1668: Summary Report of the Voyage Made to the Californias by Captain Francisco De Lucenilla*, trans. W. Michael Mathes (Los Angeles, CA, 1966)

Chadour-Sampson, Beatriz, and Hubert Bari, *Pearls* (London, 2013)

Clark, Grahame, *Symbols of Excellence: Precious Materials as Expressions of Status* (New York, 1986)

Clunas, Craig, *Superfluous Things: Material Culture and Social Status in Early Modern China* (Urbana, IL, 1991)

Content, Derek, ed., *The Pearl and the Dragon: A Study of Vietnamese Pearls and a History of the Oriental Pearl Trade* (Houlton, ME, 1999)

Crosby, Harry W., *Californio Portraits: Baja California's Vanishing Culture* (Norman, OK, 2015)

Dakin, W. J., *Pearls* (Cambridge, 1913)

Dawson, Kevin, *Undercurrents of Power: Aquatic Culture in the African Diaspora* (Philadelphia, PA, 2018)

De Romanis, Federico, and Marco Maiuro, eds, *Across the Ocean: Nine Essays on Indo-Mediterranean Trade* (Boston, MA, 2015)

Dirlam, Dona Mary, and Robert Weldon, eds, *Splendour and Science of Pearls* (Carlsbad, CA, 2013)

Donkin, R. A., *Beyond Price: Pearls and Pearl Fishing, Origins to the Age of Discoveries* (Philadelphia, PA, 1998)

Dubin, Lois Sherr, *The History of Beads from 100,000 BC to the Present* (New York, 2009)

Edwards, Hugh, *Port of Pearls: A History of Broome* (Swanbourne, WA, 1988)

Edwards, Tanya, and Sarah Yu, *Lustre: Pearling and Australia* (Welshpool, Western Australia, 2018)

Ericson, Kjell David, 'Nature's Helper: Mikimoto Kōkichi and the Place of Cultivation in the Twentieth Century's Pearl Empires', PhD dissertation, Princeton University, 2015

Eunson, Robert, *The Pearl King: The Story of the Fabulous Mikimoto* (Tokyo, 1964)

Finlay, Victoria, *Jewels: A Secret History* (New York, 2006)

Foulkes, Nick, *Mikimoto* (New York, 2008)

Frankopan, Peter, *The Silk Roads: A New History of the World* (New York, 2015)

Frick, Carole Collier, *Dressing Renaissance Florence: Families, Fortunes and Fine Clothing* (Baltimore, MA, 2002)

Gardner, Andrew M., *City of Strangers: Gulf Migration and the Indian Community in Bahrain* (Ithaca, NY, 2010)

Gems and Gemology, the quarterly scientific journal of the Gemological Institute of America

Gervis, M. H., and N. A. Sims, *The Biology and Culture of Pearl Oysters* (London, 1992)

Gillis, John, and Franziska Torma, eds, *Fluid Frontiers: New Currents in Marine Environmental History* (Cambridge, 2015)

Goggin, Maureen Daly, and Beth Fowkes Tobin, eds, *Women and the Material Culture of Death* (Abingdon, 2014)

Graz, Marie-Christine Autin, *Jewels in Painting* (Milan, 1999)

Hackney, Ki, and Diana Edkins, *People and Pearls: The Magic Endures* (New York, 2000)

Hardy, Robert, *Travels in the Interior of Mexico in 1825, 1827 and 1828* (London, 1829)

Hayashi, Ryoichi, *The Silk Road and the Shoso-in*, trans. Robert Ricketts, The Heibonsha Survey of Japanese Art #6 (New York, 1975)

Hellyer, Peter, *Waves of Time: The Maritime History of the United Arab Emirates* (London, 1998)

Hertz, Bram, *A Catalogue of the Collection of Pearls and Precious Stones Formed by Henry Philip Hope, Esq.* (London, 1839)

Hornell, James, *Report of the Government of Madras on the Indian Pearl Fisheries in the Gulf of Mannar* (Madras, 1905)

Humphreys, Christmas, *The Great Pearl Robbery of 1913: A Record of Fact* (London, 1929)

Ibn Battutah, *Travels in Asia and Africa, 1325–1354*, trans. H.A.R. Gibb (London, 1929)

Insoll, Timothy, *The Land of Enki in the Islamic Era: Pearls, Palms, and Religious Identity in Bahrain* (London, 2005)

Jeffries, David, *A Treatise on Diamonds and Pearls* (London, 1751)

Jones, Prudence, *Cleopatra: A Sourcebook* (Norman, OK, 2006)

Joyce, Kristin, and Shellei Addison, *Pearls: Ornament and Obsession* (New York, 1993)

Karampelas, Stefanos, et al., *Gems and Gemmology: An Introduction for Archaeologists, Art-Historians and Conservators* (New York, 2020)

Komaroff, Linda, et al., *Gifts of the Sultan: The Arts of Giving at the Islamic Courts* (New Haven, CT, 2011)

Kornitzer, Louis, *The Pearl Trader* (New York, 1937)

Kunz, George Frederick, 'The Exhibition of Pearls at the World's Columbian Exposition', *Bulletin of the United States Fish Commission*, XIII (Washington, DC, 1893)

——, *Gems and Precious Stones of North America* (New York, 1890)

Kunz, George Frederick, and Charles Hugh Stevenson, *The Book of the Pearl: The History, Art, Science, and Industry of the Queen of Gems* (New York, 1908)

Landman, Neil, et al., *Pearls: A Natural History* (New York, 2001)

Lanfranconi, Claudia, *Girls in Pearls: The Story of a Passion in Paintings and Photographs* (New York, 2006)

Las Casas, Bartolomé de, *A Short Account of the Destruction of the Indies* [1542], ed. and trans. Nigel Griffin, with an introduction by Anthony Pagden (London, 2004)

Le Comte, Louis, *Memoirs and Observations Typographical, Physical, Mathematical, Mechanical, Natural, Civil, and Ecclesiastical, Made in a Late Journey through the Empire of China* (London, 1699)

Leonardi, Camillo, *The Mirror of Stones* [1502] (London, 1750)

Linnaeus, Carl, *Lachesis Lapponica: A Tour in Lapland*, ed. James Edward Smith (London, 1811)

Lintilhac, Jean-Paul, *Black Pearls of Tahiti* (Papeete, Tahiti, 1987)

Lorimer, John Gordon, *Gazetteer of the Persian Gulf, Oman and Central Arabia* (Calcutta, 1908, 1915)

Loring, John, *Tiffany Pearls* (New York, 2006)

Macdonald, Alexander, *In the Land of Pearl and Gold: A Pioneer's Wanderings in the Backblocks and Pearling Grounds of Australia and New Guinea* (London, 1907)

Machado, Pedro, et al., eds, *Pearls, People and Power: Pearling and Indian Ocean Worlds* (Athens, OH, 2019)

Mackintosh-Smith, Tim, *Travels with a Tangerine: A Journey in the Footnotes of Ibn Battutah* (London, 2001)

Malaguzzi, Silvia, *The Pearl* (New York, 2001)

Maraini, Fosco, *The Island of the Fisherwomen* [1960], trans. Eric Mosbacher (New York, 1962)

Markel, Stephen, ed., *India's Fabled City: The Art of Courtly Lucknow* (Los Angeles, CA, 2011)

Martinez, Delores, *Identity and Ritual in a Japanese Diving Village* (Honolulu, HI, 2004)

Martínez, Julia, and Adrian Vickers, *The Pearl Frontier: Indonesian Labor and Indigenous Encounters in Australia's Northern Trading Network* (Honolulu, HI, 2015)

Muller, Priscilla, *Jewels in Spain, 1500–1800* (Madrid, 2012)

Mullins, Stephen, *Octopus Crowd: Maritime History and the Business of Australian Pearling in Its Schooner Age* (Tuscaloosa, AL, 2019)

—, *Torres Strait: A History of Colonial Occupation and Culture Contact, 1864–1897* (Rockhampton, Queensland, 1995)

Museo Universitario de Arte Contemporáneo, *Fritzia Irízar: Mazatlanica* (Mexico City, 2019)

Nadelhoffer, Hans, *Cartier: Jewelers Extraordinary* (New York, 1984)

Neat, Timothy, ed., *The Summer Walkers: Travelling People and Pearl-Fishers in the Highlands of Scotland* (Edinburgh, 2002)

Newbury, Colin, *Tahiti Nui: Change and Survival in French Polynesia, 1767–1945* (Honolulu, HI, 1980)

Newman, Renee, *Pearl Buying Guide: How to Evaluate, Identify and Select Pearls and Pearl Jewelry* (Los Angeles, CA, 1992)

Nicols, Thomas, *A Lapidary; or, The History of Pretious Stones* (Cambridge, 1652)

Opie, Catherine, et al., *700 Nimes Road* (New York, 2015)

Palgrave, William Gifford, *Personal Narrative of a Year's Journey Through Central and Eastern Arabia, 1862–63* (London, 1883)

Pearl, trans. Simon Armitage (New York, 2016)

Pliny the Elder, *The Natural History*, ed. John Bostock and H. T. Riley, in the Perseus Digital Library, www.perseus.tufts.edu

Pointon, Marcia, *Brilliant Effects: A Cultural History of Gem Stones and Jewellery* (London, 2009)

Raden, Aja, *Stoned: Jewelry, Obsession, and how Desire Shapes the World* (New York, 2015)

Rees, Coralie, and Leslie Rees, *Coasts of Cape York: Travels Around Australia's Pearl-Tipped Peninsula* (Sydney, 1960)

Reynolds, Henry, *North of Capricorn: The Untold Story of Australia's North* (Sydney, 2003)

Ribeiro, João, *The Historic Tragedy of the Island of Ceilão*, trans. P. E. Pieris (New Delhi, 1999)

Rosenthal, Léonard, *The Kingdom of the Pearl* (London, 1920)

—, *The Pearl Hunter: An Autobiography* (New York, 1952)

Saville-Kent, William, *The Great Barrier Reef of Australia: Its Products and Potentialities* (London, 1893)

Scales, Helen, *Spirals in Time: The Secret Life and Curious Afterlife of Seashells* (London, 2015)

Scarisbrick, Diana, *Ancestral Jewels* (London, 1989)

—, Christophe Vachaudez and Jan Walgrave, eds, *Brilliant Europe: Jewels from European Courts* (Brussels, 2008)

Screech, Timon, *Sex and the Floating World: Erotic Images in Japan, 1700–1820* (Honolulu, HI, 1999)

Seetah, Krish, ed., *Connecting Continents: Archaeology and History in the Indian Ocean World* (Athens, OH, 2018)

Shamlan, Sayf Marzuq, *Pearling in the Arabian Gulf: A Kuwaiti Memoir* (London, 2000)

Shirai, Shohei, *The Story of Pearls* (Tokyo, 1970)

Sivasundaram, Sujit, *Islanded: Britain, Sri Lanka, and the Bounds of an Indian Ocean Colony* (Chicago, IL, 2013)

Skinner, Ann, Mark Young and Lee Hastie, *Ecology of the Freshwater Pearl Mussel*, Conserving Natura 2000 Rivers, Ecology Series, no. 2 (Edinburgh, 2003)

Smaal, Aad C., et al., eds, *Goods and Services of Marine Bivalves* (Cham, 2019)

Southgate, Paul C., and John S. Lucas, eds, *The Pearl Oyster* (Amsterdam, 2008)

Steinbeck, John, *The Log from the Sea of Cortez* [1951] (London, 1986)
——, *The Pearl* [1947] (New York, 1992)
Steinberg, Philip E., *The Social Construction of the Ocean* (Cambridge, 2001)
Steuart, James, *An Account of the Pearl Fisheries of Ceylon* (n.p., 1843)
Strack, Elisabeth, *Pearls* (Stuttgart, 2006)
Streeter, Edwin W., *Pearls and Pearling Life* (London, 1886)
Tavernier, Jean-Baptiste, *Travels in India*, trans. V. Ball, 2 vols (London, 1889)
Taylor, Elizabeth, *My Love Affair with Jewelry* (London, 2003)
Thompson, J. Eric S., ed., *Thomas Gage's Travels in the New World*
    (Westport, CT, 1981)
Trethewey, Rachel, *Pearls before Poppies: The Story of the Red Cross Pearls*
    (Stroud, 2018)
Tuson, Penelope, ed., *The Persian Gulf Trade Reports, 1905–1940*
    (Cambridge, 1987)
Urkevich, Lisa, *Music and Traditions of the Arabian Peninsula: Saudi Arabia,
    Kuwait, Bahrain, and Qatar* (New York, 2015)
Vertrees, Herbert, *Pearls and Pearling* (New York, 1913)
Villiers, Alan, *Sons of Sinbad* (New York, 1940)
Walker, Susan, ed., *Ancient Faces: Mummy Portraits from Roman Egypt*
    (New York, 2000)
Walther, Michael, *Pearls of Pearl Harbor and the Islands of Hawaii: The History,
    Mythology, and Cultivation of Hawaiian Pearls* (Honolulu, HI, 1997)
Ward, Fred, *Pearls* (Bethesda, MD, 1998)
Warsh, Molly, *American Baroque: Pearls and the Nature of Empire, 1492–1700*
    (Chapel Hill, NC, 2018)
Wise, Richard, *Secrets of the Gem Trade: The Connoisseur's Guide to Precious
    Gemstones* (Lenox, MA, 2004)
Woodward, Fred, *The Scottish Pearl in Its World Context* (Edinburgh, 1994)
Wright, Elaine, et al., *Muraqqa': Imperial Mughal Albums from the Chester
    Beatty Library, Dublin* (Alexandria, VA, 2008)

Acknowledgements

A pearl is a small thing, but my debts are large. Living in southern California, I have been fortunate to have access to three major research libraries – the Richard T. Liddicoat Gemological Library at the Gemological Institute of America, The Huntington Library and The Getty Research Institute library. I would like to thank the staff at these institutions, particularly Marie Svoboda, associate conservator in the Department of Antiquities Conservation at the J. Paul Getty Museum, and Rose Tozer, Senior Research Librarian at GIA. I am also indebted to the ever-helpful librarians at my own institution, Chapman University, who supplied me with books during the long stretches of pandemic lockdowns, especially Justine Lim and Lien T. N. Nguyen. Chapman colleagues Patrick Fuery, Wendy Salmond, Natalie Lawler and Jessica Bocinski lent much appreciated support. I am especially grateful to Tom Zoellner, who read part of this manuscript, and to the MFA students in his Creative Non-fiction class for lively discussions on the Mikimoto Shanghai pearl heist. I also thank Bill Wright at Chapman, Edgar T. Walters, University of Texas, Houston, and Robyn Crook, San Francisco State University, for thoughtfully answering my questions on molluscs' experience of pain and the ethics of periculture.

Sumanth Prabhaker, editor in chief of *Orion*, provided invaluable feedback on my writing on pearls in Scotland and I am grateful to *Orion* for the opportunity to publish a personal exploration and include excerpts in this book. I am also grateful to the *Washington Post* for permission to include excerpts from my 2020 article on black pearls and politicians and to editor Carly Goodman and writer Stephen G. Bloom for their helpful advice.

In Japan, I was fortunate to meet Christopher Douglas of Mikimoto Pearl Island and I thank him for his kind assistance and copies of archival photographs. I extend thanks too to Ryotaro Ozaki of Ehime University. In my home country of Scotland, Audrey Macbeth and the Buchanan family supplied encouragement and photography. I am indebted to ecologist Peter Cosgrove for discussions and literature on the European pearl mussel, and to museum colleagues David Forsyth and Margaret Wilson of National Museums Scotland, Lauren Gill of the Universiy of Glasgow, and scholar Michael Pearce. Many museum colleagues were generous with their time and resources, and I would like to thank Roger Portell of the Florida Museum of Natural History for his image of fossil pearls.

The artist Richard Turner was a constant source of stimulating conversation about the timeliness of my research in relation to contemporary artists' engagement with this subject. I am thankful to Richard for introducing me to the work of Mexican artist Fritzia Irízar, and to the artist for generously sharing images of her work. I acknowledge artist Ivan Forde for a stimulating conversation on the *Epic of Gilgamesh*, and Alisson Gothz for sharing her portrait of an artistic life. I am grateful to Catherine Opie for allowing me to reproduce her superb portrait of the jewel La Peregrina (Elizabeth Taylor), and to the jewellers Sarah Ho, Pernille Lauridsen and Nadine Leo. I thank Stacey Kurzendofer for her leads on the revived pearling tradition in Kuwait.

Throughout the course of my research, I was fortunate to draw on the expertise of many industry experts. I am especially thankful to Betty Sue King, Anil Maloo of Baggins Pearls, Devchand Chodhry of Orient Pearl, David Blackman and BJ Curtis. I thank, too, Laurent E. Cartier, co-founder of the invaluable Sustainable Pearls project, for his generosity in providing superb images of Kamoka Pearl Farm, French Polynesia.

Veli-Pekka Lehtola, Heini Wesslin, Jonas Monié Nordin, Gunlög Fur and Jan Asplund patiently answered my questions about pearls in Swedish Lapland.

I was the fortunate recipient of a Winterthur Research Fellowship and had on-site access to American portraiture depicting subjects in pearls. I am grateful to Ann Wagner, Chase Markee, Carley Altenburger, Sarah Lewis and Laura Fravel for making my visit productive and memorable, and to my fellow Scholars-in-Residence Malcolm Yiyun Huang, Gloria and Sue Johnson for making it superbly enjoyable. I am deeply thankful to Judy and Bruce Hoechner for hosting me in Philadelphia.

I am fortunate to publish once more with Reaktion Books and am grateful to Michael Leaman for giving me this opportunity again to work with his talented, creative staff. I would like to thank Amy Salter, Alex Ciobanu, Susannah Jayes, Maria Kilcoyne and proofreader Emma Wiggin for taking such care with this publication.

Lastly, I thank beloved friends Kathleen Campbell, Irena Farrell, Laura Leir, Christine Linnell, Timothy G. Fleming, Christina Benedict-Young and Katy Gow for encouraging my love of these small, water-born gems; my sons, Ewan and Owen, who might not have understood the obsession but allowed me moments to write, and my husband, Byron Shen, for, among many other gestures of love, gifting me with strands of pearls.

✿

Photo

Acknowledgements

The author and publishers wish to thank the organizations and individuals listed below for authorizing reproduction of their work.

British Library, London: pp. 63, 170; images courtesy Laurent E. Cartier: pp. 49, 50, 51; © The Trustees of the Chester Beatty Library, Dublin, CBL in 11A.69: pp. 232, 233; The Art Institute of Chicago: pp. 62, 140; Cooper Hewitt, Smithsonian Design Museum: pp. 56 (museum purchase through gift of various donors), 105 (gift of anonymous donor), 126 bottom (museum purchase from Smithsonian Institution Collections Acquisition Program, Decorative Arts Association Acquisition, and Sarah Cooper-Hewitt Funds), 224 (gift of the Estate of James Hazen Hyde); Mary Evans Picture Library: p. 197 (Illustrated London News Ltd); Florida Museum, Invertebrate Paleontology Collection: p. 18 (image courtesy Roger W. Portell); the J. Paul Getty Museum, Los Angeles: p. 178 (gift of Dr Louis F. D'Elia, IV and Michael D. Salazar, © Robert Neal Stivers) and pp. 65 and 221 (digital images courtesy of the Getty's Open Content Program); the J. Paul Getty Museum, Villa Collection, Malibu, California: pp. 150 and 213 (digital images courtesy of the Getty's Open Content Program), 217 (photograph by Marie Svoboda, image © 2017 J. Paul Getty Trust); photograph courtesy Alisson Gothz: p. 198; Sarah Ho Jewellery: p. 44; The Huntington Art Museum and Botanical Gardens: p. 9; *Illustrated London News*, 14 May 1921: p. 117; photographs courtesy Fritzia Irízar: p. 176 top and bottom; *Japan Times*: p. 112; photographs courtesy Betty Sue King: pp. 35, 38, 168; photograph courtesy Pernille Lauridsen: p. 159; photograph courtesy Nadine Ifrah Leo, the Mississippi River Pearl Jewelry Co.: p. 26; Library of Congress, Washington, DC: pp. 17 top and bottom left; Library of Congress, Geography and Map Division: p. 167; permission of the Linnean Society of London: p. 101; Los Angeles County Museum of Art: pp. 237 top (gift of Mr and Mrs H. K. Lee), 237 bottom (purchased with funds provided by the Art Museum Council in honour of the museum's 40th anniversary through the 2005 Collectors Committee); photographs courtesy Anil Maloo, Baggins Pearls, Los Angeles: pp. 6, 133; Mauritshuis, The Hague, Netherlands: pp. 76, 80, 222, 227; The Metropolitan Museum of Art, New York: pp. 14 (Robert Lehman Collection, 1975), 36 (gift of J. Pierpont Morgan, 1917), 78 (gift of J. Pierpont Morgan, 1917), 156 (Rogers Fund, 1914); 181 (Purchase, Acquisitions Fund, Christopher C. Grisanti and Suzanne P. Fawbush, Austin B. Chinn, and Katharine R. Brown gifts, gifts of Marx Freres, J. Pierpont Morgan, and Mrs Frank D. Millett, by exchange, and funds from various donors, 2007), 189 (Purchase, Mr and Mrs Claus von Bülow gift, 1978), 192 (bequest of Mrs Charles Wrightsman, 2019), 200 (Purchase, Rogers Fund and the Kevorkian Foundation gift, 1955), 203 (Purchase, Rogers Fund and The Kevorkian Foundation gift, 1955), 214 (Rogers Fund, 1920); Mikimoto Pearl Island Museum Archives: pp. 103, 107, 108, 121 bottom (photographs courtesy Christopher Douglas); Nationaal Archief, The Hague, Netherlands: p. 118 top; National Archives and Records Administration (NARA), Washington, DC: pp. 68, 69; © The National Gallery, London: p. 231; National Gallery of Victoria, Melbourne: p. 145 (Felton bequest, 1933); Copyright © National Land Image Information (Colour Aerial Photographs), Ministry of Land, Infrastructure, Transport and Tourism: p. 109; National Museum of Natural History, Smithsonian Open Access: p. 90 top and bottom; National Museums Scotland: pp. 81, 84; National Portrait Gallery, London: pp. 75, 182; NOAA/OER (Bioluminescence 2009 Expedition): p. 23 top; NOAA Photo Library: pp. 17 bottom right (photograph by Dr Dwayne Meadows), 171 (archival photograph by Stefan Claesson, Gulf of Maine Cod Project, NOAA National Marine Sanctuaries; courtesy of National Archives); *North-China Desk Hong List*, July 1917: p. 206; © Catherine Opie: p. 240 (courtesy Regen Projects, Los Angeles and Lehmann Maupin, New York, Hong Kong and Seoul); image courtesy Ryotaro Ozaki: p. 30; Rijksmuseum, Amsterdam, Netherlands: pp. 27, 77, 86, 162 (on loan from the Cultural Heritage Agency of the Netherlands, Amersfoort); David Rumsey Historical Map Collection: pp. 52, 70; by permission, Hendrik Schicke: p. 20; Fiona Shen: pp. 21, 22 top, 23 bottom, 24, 47, 113, 143 top and bottom; Fiona Shen, Mikimoto Pearl Island: pp. 31, 116; photograph by Donald Shields: p. 83 (courtesy of Peter Cosgrove); Smithsonian American Art Museum Collection: pp. 82, 191 (bequest of Harriet Lane Johnston); collection of the Smithsonian National Museum of African American History and Culture: p. 13; Smithsonian National Museum of Asian Art: p. 139 (Katsushika Hokusai/Freer Gallery of Art Study Collection, Smithsonian Institution, Washington, DC: Purchase, The Gerhard Pulverer Collection – Charles Lang Freer Endowment, Friends of the Freer and Sackler Galleries and the Harold P. Stern Memorial Fund in appreciation of Jeffrey P. Cunard and his exemplary service to the Galleries as chair of the Board of Trustees (2003–7), FSC-GR-780.4.3); *The Sphere*, 21 May 1921: p. 121 top; State Library of Queensland (via Flickr): pp. 127, 128, 129; Victoria and Albert Museum, London: pp. 40, 64, 158, 161, 164; The Walters Art Museum, Baltimore: pp. 37 (acquired by Henry Walters), 43 top and bottom (gift of Miss Laura F. Delano, 1977), 66 (acquired by Henry Walters, 1904), 97 (acquired by Henry Walters), 190 (acquired by Henry Walters), 218 (acquired by

Index